T0225729

Satoyama Initiative Thematic Review

Series Editor

Tsunao Watanabe, United Nations University Institute for the Advanced Study of Sustainability (UNU-IAS), Tokyo, Japan

This Open Access book series aims to make timely and targeted contributions for decision-makers and on-the-ground practitioners by producing knowledge concerning "socio-ecological production landscapes and seascapes" (SEPLS) – areas where production activities help maintain biodiversity and ecosystem services in various forms while sustainably supporting the livelihoods and well-being of local communities. Each volume will be designed as a compilation of case studies providing useful knowledge and lessons focusing on a specific theme that is important for SEPLS, accompanied by a synthesis chapter that extracts lessons learned through the case studies to present them for policy-relevant academic discussions. The series is also intended to contribute to efforts being made by researchers and scientists to strengthen the evidence base on socio-ecological dynamics and resilience, including those under the Intergovernmental Science-Policy Platform on Biodiversity and Ecosystem Services (IPBES) and the Convention on Biological Diversity (CBD).

The promotion and conservation of SEPLS have been the focus of the Satoyama Initiative, a global effort to realise societies in harmony with nature. In 2010, the International Partnership for the Satoyama Initiative (IPSI) was established to implement the concept of the Satoyama Initiative and promote various activities by enhancing awareness and creating synergies among those working with SEPLS. As a unique platform joined by governmental, intergovernmental, nongovernmental, private-sector, academic, and indigenous peoples' organisations, IPSI has been collecting and sharing the information, lessons, and experiences on SEPLS to accumulate a diverse range of knowledge. Case studies to be published in this series will be solicited from practitioners, researchers, and others working on the ground, who are affiliated with IPSI member organisations and closely involved and familiar with the activities related to SEPLS management.

Maiko Nishi • Suneetha M. Subramanian

Editors

Ecosystem Restoration through Managing Socio-Ecological Production Landscapes and Seascapes (SEPLS)

 Springer

Editors
Maiko Nishi
Biodiversity & Society
United Nations University Institute for the
Advanced Study of Sustainability
(UNU-IAS)
Tokyo, Japan

Suneetha M. Subramanian
Biodiversity & Society
United Nations University Institute for the
Advanced Study of Sustainability (UNU-IAS)
Tokyo, Japan

This work was supported by the Ministry of the Environment, Japan (MoEJ)

ISSN 2731-5169   ISSN 2731-5177   (electronic)
Satoyama Initiative Thematic Review
ISBN 978-981-99-1294-0   ISBN 978-981-99-1292-6   (eBook)
https://doi.org/10.1007/978-981-99-1292-6

© UNU-IAS 2023. This book is an open access publication.
The opinions expressed in this publication are those of the authors/editors and do not necessarily reflect the
views of UNU-IAS, its Board of Directors, or the countries they represent.
Open Access This book is licenced under the terms of the Creative Commons Attribution-
NonCommercial-ShareAlike 3.0 IGO licence (http://creativecommons.org/licenses/by-nc-sa/3.0/igo/),
which permits any noncommercial use, sharing, adaptation, distribution, and reproduction in any
medium or format, as long as you give appropriate credit to UNU-IAS, provide a link to the Creative
Commons licence and indicate if changes were made. If you remix, transform, or build upon this book or a
part thereof, you must distribute your contributions under the same licence as the original.
The use of the UNU-IAS name and logo, shall be subject to a separate written licence agreement between
UNU-IAS and the user and is not authorised as part of this CC BY-NC-SA 3.0 IGO licence. Note that the
link provided above includes additional terms and conditions of the licence.
The images or other third party material in this book are included in the book's Creative Commons licence,
unless indicated otherwise in a credit line to the material. If material is not included in the book's Creative
Commons licence and your intended use is not permitted by statutory regulation, or exceeds the permitted
use, you will need to obtain permission directly from the copyright holder.
This work is subject to copyright. All commercial rights are reserved by the author(s), whether the whole
or part of the material is concerned, specifically the rights of translation, reprinting, reuse of illustrations,
recitation, broadcasting, reproduction on microfilms or in any other physical way, and transmission or
information storage and retrieval, electronic adaptation, computer software, or by similar or dissimilar
methodology now known or hereafter developed. Regarding these commercial rights a non-exclusive
licence has been granted to the publisher.
The use of general descriptive names, registered names, trademarks, service marks, etc. in this publication
does not imply, even in the absence of a specific statement, that such names are exempt from the relevant
protective laws and regulations and therefore free for general use.
The publisher, the authors, and the editors are safe to assume that the advice and information in this book
are believed to be true and accurate at the date of publication. Neither the publisher nor the authors or the
editors give a warranty, expressed or implied, with respect to the material contained herein or for any
errors or omissions that may have been made. The publisher remains neutral with regard to jurisdictional
claims in published maps and institutional affiliations.

This Springer imprint is published by the registered company Springer Nature Singapore Pte Ltd.
The registered company address is: 152 Beach Road, #21-01/04 Gateway East, Singapore 189721,
Singapore

Foreword

Humanity is going through increasingly challenging times as we face a confluence of environmental degradation, biodiversity loss, and climate change. Each of these crises can be traced back to human action and decisions related to the use of land and natural resources, consumption patterns, and equity in transactions between people. The state of environmental degradation and its negative impacts on human well-being have received increasing recognition in recent years, with policy statements and commitments made in global forums highlighting the urgency of the issue. In the last few years, high-profile intergovernmental assessments have highlighted the drivers of ecosystem degradation and possible pathways for recovery. The challenge now is to translate these objectives into action on the ground.

In this context it is critical to identify, share, and scale up successful approaches. This publication aims to further this process, by compiling case studies on the theme of ecosystem restoration learning from socio-ecological production landscapes and seascapes (SEPLS). It explores and draws lessons on how to prevent loss and degradation of ecosystems and ensure their recovery at the ground level, based on the experiences of members of the International Partnership for the Satoyama Initiative (IPSI). These detail a variety of contexts and activities, illustrating how bottom-up approaches can contribute to achieving policy objectives. They also highlight the need to address asymmetries in capacities across stakeholder groups, and foster a systemic approach in decision-making and implementation.

The United Nations University (UNU) Institute for the Advanced Study of Sustainability (UNU-IAS) is focused on mobilising knowledge concerning environment, development, and sustainability to foster solutions. Academic publications such as this volume are vital tools to synthesise and transfer knowledge from diverse experience on the ground, to inform and improve policymaking. UNU recently joined the UN Decade on Ecosystem Restoration, and our research is advancing this global partnership to protect and revive ecosystems for the benefit of both people and nature. As the international community has recently developed the next global framework for biodiversity, I believe readers will find this volume not only engaging and informative, but also highly useful and practical.

United Nations University   Shinobu Yume Yamaguchi
Institute for the Advanced Study of
Sustainability
Tokyo, Japan

Foreword

Ecosystem restoration is indispensable to sustain our lives and pursue sustainable development. There is irrefutable evidence that biodiversity loss and land and sea degradation are adversely affecting the ability of ecosystems to function well. This implies that the support systems for all life are negatively impacted, and the well-being of all humankind is eroding. Ironically, the COVID-19 pandemic was a wake-up call to all decision makers that for humanity to flourish, we should *live in harmony with nature*. This has been the refrain of the work of the Satoyama Initiative since it was conceived more than a decade ago.

The initiative and its members through the International Partnership for the Satoyama Initiative (IPSI) have constantly striven to highlight how harmonious human-nature relations are expressed through various activities in socio-ecological production landscapes and seascapes (SEPLS). These on-the-ground activities hold pragmatic lessons of policy relevance and for further replication in similar contexts. They are bound by certain principles that cut across implementation scales: a diversity of land and sea uses is encouraged, multi-stakeholder engagement in an inclusive and equitable manner is the preferred mode of decision-making, and harvesting multiple benefits from the various activities is expected (from conservation, food and health security, climate adaptation, acknowledgement and use of different types of knowledge, recognition of spiritual, cultural and other intangible priorities, among others). This then traverses sectoral boundaries and the priorities of different actors in a landscape on the land and sea use.

These experiences inform us of challenges and opportunities on the ground to implement various social and ecological policy objectives. Since 2015, we have been compiling the experiences of IPSI members towards particular policy objectives and publishing them as the Satoyama Initiative Thematic Review (SITR). We intentionally keep the language of the volume simple to understand in order to appeal to both specialists and practitioners. This is the eighth edition of SITR and the third published in collaboration with Springer Nature. In this volume, we focus on the theme of ecosystem restoration in the context of SEPLS. The synthesis chapter collates key messages from across all the other chapters to give a bird's eye view of the issues involved, including the drivers of ecosystem degradation and loss, and good practices, opportunities, and incentives to address them. I hope you enjoy reading the volume as much as we enjoyed putting it together. We also hope very much that it is the first of many contributions we can make to the ongoing efforts under the United Nations Decade on Ecosystem Restoration to raise awareness, develop the capacities of different stakeholders on tackling challenges, leverage opportunities, and highlight best practices on the theme.

United Nations University   Tsunao Watanabe
Institute for the Advanced Study of
Sustainability
Tokyo, Japan

The Secretariat of the International
Partnership for the Satoyama Initiative
Tokyo, Japan

Preface

The Satoyama Initiative is "a global effort to realise societies in harmony with nature", started through a collaboration between the United Nations University (UNU) and the Ministry of the Environment of Japan. The initiative focuses on the revitalisation and management of "socio-ecological production landscapes and seascapes" (SEPLS), areas where production activities help maintain biodiversity and ecosystem services in various forms while sustainably supporting the livelihoods and well-being of local communities. In 2010, the International Partnership for the Satoyama Initiative (IPSI) was established to implement the concept of the Satoyama Initiative and promote various activities by enhancing awareness and creating synergies among those working with SEPLS. IPSI provides a unique platform for organisations to exchange views and experiences and to find partners for collaboration. As of April 2023, 298 members have joined the partnership, including governmental, intergovernmental, non-governmental, private sector, academic, and indigenous peoples' organisations.

The Satoyama Initiative promotes the concept of SEPLS through a threefold approach that argues for connection of land- and seascapes holistically for management of SEPLS (see Fig. 1). This often means involvement of various sectors at the landscape or seascape scale, under which it seeks to: (1) consolidate wisdom in securing diverse ecosystem services and values, (2) integrate traditional ecosystem knowledge and modern science, and (3) explore new forms of co-management systems. Furthermore, activities for SEPLS management cover multiple dimensions, such as equity, addressing poverty and deforestation, and incorporation of traditional knowledge for sustainable management practices in primary production processes such as agriculture, fisheries, and forestry (UNU-IAS and IGES 2015).

As one of its core functions, IPSI serves as a knowledge-sharing platform through the collection and sharing of information and experiences on SEPLS, providing a place for discussion among members and beyond. Over 280 case studies have been collected and are shared on the IPSI website, providing a wide range of knowledge covering diverse issues related to SEPLS. Discussions have also been held to further strengthen IPSI's knowledge-facilitation functions, with members suggesting that

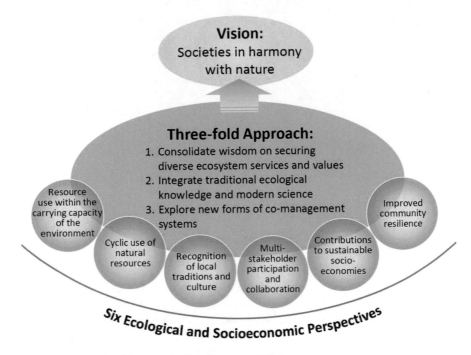

Fig. 1 The conceptual framework of the Satoyama Initiative

efforts should be made to produce knowledge on specific issues in SEPLS in order to make more targeted contributions to decision makers and on-the-ground practitioners.

It is in this context that a project to create a publication series titled the "Satoyama Initiative Thematic Review" (SITR)[1] was initiated in 2015. The SITR series was developed as a compilation of case studies providing useful knowledge and lessons focusing on a specific theme that is important for SEPLS. The overall aim of the SITR publications is to collect experiences and relevant knowledge, especially from practitioners working on the ground, considering their usefulness in providing concrete and practical knowledge and information as well as their potential to contribute to policy recommendations. Each volume is also accompanied by a synthesis chapter which extracts lessons learned through the case studies, presenting them for policy-relevant academic discussions. This series also contributes to efforts being made by researchers to strengthen the evidence base for policymaking concerning social-ecological dynamics and resilience, including those under the Intergovernmental Science-Policy Platform on Biodiversity and Ecosystem Services (IPBES) and the Convention on Biological Diversity (CBD). Seven volumes have been published since 2015 for this series on various timely topics such as knowledge

---

[1] The previous volumes of the SITR series are available at: https://satoyama-initiative.org/featured_activities/sitr/

enhancement, biodiversity mainstreaming, sustainable livelihoods, effective area-based conservation, multiple values associated with sustainable use of biodiversity, transformative change, and nexus between biodiversity, health and sustainability.

Building on the earlier volumes that were in-house publications of UNU, the sixth edition and onward have become published by Springer Nature to reach out to a broader range of readers. In particular, the SITR has become a Springer book series from the seventh volume in an aim to enhance consistent and coherent contributions to science-policy-practice interfaces, while maintaining the publications' impact and reach to a wide audience. Furthermore, a review committee has become established since the seventh volume to engage in the review process for publication. By inviting experienced and knowledgeable experts in SEPLS management from the IPSI community, this expert group has helped to ensure credibility and reinforce the quality of the contents and at the same time to facilitate sharing of expertise and collaboration among IPSI members.

Similar to the previous volumes, this eighth edition was developed through a multi-stage process, including both peer review and discussion among the authors and reviewers at a workshop. Authors had several opportunities to receive feedback, which helped them to improve their manuscripts in substance, quality, and relevance. First, each manuscript received comments from the editorial team and the review committee relating primarily to its contributions to the theme of the volume. Peer review was then conducted by authors of other chapters. Each author received feedback from two other authors who were requested to comment on whether the manuscript was easy to understand and informative, addressed key questions of the volume's focus, and provided useful lessons. The aforementioned workshop was then held virtually on 30 May–1 June 2022 to enable the exchange of feedback between authors and reviewers. The basic ideas contained in the synthesis of the concluding chapter were developed from presentations and discussions during the workshop, and the chapter was made available for review by authors and reviewers before finalisation.

The above process offers an opportunity for authors from both academic and non-academic organisations to contribute to generating knowledge in an accessible and interactive way, as well as to provide high-quality papers written in simple language for academics and a broader audience alike. It is our hope that this publication will be useful in providing information and insights to practitioners, researchers, and policymakers on the importance of long-term collaborative management of SEPLS for preventing, halting, and reversing land and sea degradation, contributing to ecosystem restoration and sustainable development. This, we hope, will prompt policymaking that strengthens such integrated and holistic management approaches.

We would like to thank all the authors who contributed their case studies. We also appreciate the continued commitment and support for the thorough review process by the four experts in the review committee: Devon Dublin, Alexandros Gasparatos, Paulina G. Karimova, and Chemuku Wekesa. We are also grateful to the other participants in the case study workshop who provided insightful remarks and valuable inputs into the discussions. These individuals include Hana Matsuzaki,

Andrea Romero, Jie Su, and Eiji Tanaka. We also thank the colleagues of UNU-IAS and Regional Environmental Planning, Inc. who were supportive and instrumental in organising the workshop and facilitating the publication process: Shoko Arao, Alexandra Franco Guajardo, Atsushi Futamura, Roberta Hübner, Gen Ito, Alebel Melaku, Bruno Leles, Tadashi Masuzawa, Yoshino Nakahara, Youichi Sonoda, Nicholas Turner, Makiko Yanagiya, Kanako Yoshino, and Madoka Yoshino. Our gratitude extends to the UNU administration, especially Francesco Foghetti, and to colleagues from UN Geospatial Information Section for all the instrumental support through the publication process. Furthermore, we acknowledge Susan Yoshimura who skilfully proofread the manuscripts.

Publication of this volume as a new Springer Book would not have been possible without the helpful guidance of Mei Hann Lee and Momoko Asawa from Springer and the institutional support and leadership of Shinobu Yume Yamaguchi and Tsunao Watanabe from UNU-IAS. Our grateful thanks are also due to the Ministry of the Environment, Japan, for supporting the activities of IPSI and its secretariat hosted by UNU-IAS.

Tokyo, Japan   Maiko Nishi
                                                                      Suneetha M. Subramanian

Reference

UNU-IAS, IGES (eds) (2015) Enhancing knowledge for better management of socio-ecological production landscapes and seascapes (SEPLS). Satoyama Initiative Thematic Review, Vol 1. United Nations University Institute for the Advanced Study of Sustainability, Tokyo

Acknowledgement

Review Committee

Devon Dublin
Alexandros Gasparatos
Paulina G. Karimova
Chemuku Wekesa

Editorial Support

Madoka Yoshino
Kanako Yoshino
Alebel Melaku
Shoko Arao
Roberta Hübner
Gen Ito
Regional Environmental Planning, Inc.

English Proofreading

Susan Yoshimura

Contents

Editors and Contributors

About the Editors

Maiko Nishi Research Fellow at United Nations University Institute for the Advanced Study of Sustainability. Her research interests include social-ecological system governance, regional planning, and agricultural land policy. PhD in Urban Planning from Columbia University.

Suneetha M. Subramanian Research Fellow at United Nations University Institute for the Advanced Study of Sustainability. Her research interests include biodiversity and human well-being with a focus on equity, traditional knowledge, community well-being, and social-ecological resilience.

About the Contributors

Archana Bhatt Works as a Scientist at MS Swaminathan Research Foundation—Community Agrobiodiversity Centre, India, on projects focused on agrobiodiversity conservation, climate change, and community development, along with training and capacity building.

Nancy Chege Works as a National Coordinator of the GEF Small Grants Programme for the United Nations Development Programme, Kenya, supporting strategic partnerships and empowerment of local communities for improved livelihoods and environmental conservation.

Jyun-Long Chen Works as a cross-discipline marine social scientist in Fisheries Research Institute, Chinese Taipei. His expertise includes economic analysis and management for marine ecosystem services, fishermen's behaviour for marine resources management, coastal resources co-management and sustainable uses.

Yufen Chuang Visual designer and Programme Specialist at Farmers' Seed Network, China. Works on projects focused on documentation and dissemination of bioculture and indigenous food systems to build rural-urban linkage and promote visualisation of sustainable agriculture and fisheries in multiple ecosystems.

C. S. Dhanya Works as a Scientist at MS Swaminathan Research Foundation—Community Agrobiodiversity Centre, India, on projects focused on agrobiodiversity conservation and livelihood enhancement of communities through technology dissemination.

Kapil Dhungana Works at the Ministry of Forest, Environment and Soil Conservation. Has worked in the soil and watershed field since 2012. Master's degree from Institute of Forestry and Fujian Agriculture and Forestry University. Passionate about biodiversity conservation and nature resource management in Nepal.

Camila I. Donatti Climate Change Adaptation Director at Conservation International. Focused on understanding the role of nature-based solutions in helping people adapt to the impacts of climate change in the Tropics, and on the monitoring and evaluation of nature-based solutions for climate adaptation.

Godwin Evenyo Dzekoto Project Manager at A Rocha Ghana, Ghana. He is interested in working with local communities to build sustainable landscapes, conserve their natural resources, and develop green value chains.

Kizito Echiru Works at Save A Seed for the Future (SAFE), Uganda, as an environment officer in projects focused on tree planting and environmental management.

Siddharth Edake Works as a Senior Manager at the World Resources Institute, India, on projects focused on forest and landscape restoration that support enhancement of national carbon sinks and are cost-effective, natural climate change solutions.

Jagger E. Enaje Fishing Regulations Officer II at the Bureau of Fisheries and Aquatic Resources. With several decades' experience in local fisheries and NGO collaboration, he is currently working in a national government agency. He works with fisherfolk on both livelihood and sustainability improvements as well as establishing marine protected areas.

Ngulefack Ernest Forghab Works as a Research Officer with the Environment and Rural Development Foundation (ERuDeF).

Gunjan Gahatraj Works at the Ministry of Forest, Environment and Soil Conservation. Has worked in the forest management field since 2012. Master of Forest Economics from Institute of Forestry and Beijing Forestry University. Interested in the economic development of the forestry sector in Nepal.

Yao-Jen Hsiao Works as an Associate Professor in the Institute of Applied Economics at National Taiwan Ocean University, on projects focused on coastal communities' environmental education and adaptive capacity.

Kang Hsu Works as a Research Assistant in Marine Fisheries Division, Fisheries Research Institute, Taiwan. The research interests focus on social-ecological system in coastal areas, stakeholder analysis, and the development of fishing villages.

Tom Kemboi Kiptenai Supports Conservation International Africa Field Division data science team in applying Earth Observation in natural resource management. Bulk of work includes mapping natural capital, and monitoring and evaluating projects by applying GIS and remote sensing.

N. Anil Kumar Works as a Senior Director at MS Swaminathan Research Foundation—Community Agrobiodiversity Centre, India, in the area of biodiversity conservation and sustainable management of genetic resources with more than three decades of experience.

Guanqi Li Programme Specialist at Farmers' Seed Network, China. Works on projects focused on farmers' seed systems enhancement to build practice-science-policy linkage and promote agrobiodiversity conservation and utilisation that enhances food security and sustainable food systems.

Chun-Pei Liao Works as a Research Assistant at Fisheries Research Institute, on projects focused on community-based marine resources management. She is also a PhD candidate in the Department of Environmental Biology and Fishery Science, National Taiwan Ocean University.

En-Yu Liu Freelance field researcher working with government agencies, scientific academies, and local NGOs in Chinese Taipei. Works on projects focused on monitoring nearshore biodiversity resources and rehabilitation projects.

Jacqueline Sapoama Mbawine Monitoring Evaluation and Reporting Manager at A Rocha Ghana, Ghana. Works on project focused on promoting sustainable woodfuel utilisation, landscape restoration, and development of green value chains for rural communities.

Alebel Melaku Research Assistant, Operating Unit Ishikawa/Kanazawa, United Nations University Institute for the Advanced Study of Sustainability. His research interests include agroforestry, climate change, biodiversity, ecosystem services, urban forestry, and urban ecology.

Tankou Christopher Mubeteneh Works as a Professor in the Department of Crop Sciences, Faculty of Agronomy and Agricultural Science, University of Dschang, Cameroon.

Yoji Natori Associate Professor at Akita International University, teaching environmental science, conservation, and sustainable development. He is also an honorary advisor to Conservation International Japan and vice chair of the Japan Committee for IUCN.

Asabaimbi Deh Nji Chief of Programmes on quality management and evaluation at the Environment and Rural Development Foundation (ERuDeF), Cameroon. He specialises in forest Audit and Certification.

Louis Nkembi Conservation and development expert. President/CEO of the Environment and Rural Development Foundation (ERuDeF).

Njukeng Jetro Nkengafac Works as Senior Research officer for the Institute of Agricultural Research for Development (IRAD) EKONA Centre and Vice President of Science, Research and Technical Services at ERuDeF.

Josephat Mukele Nyongesa Landscape Manager at Conservation International, Kenya. Works on projects focused on community-based rangeland restoration through improved livestock management and rangeland restoration to catalyse ecosystem restoration, sequester carbon, and build climate-resilient pastoral community livelihoods.

Samuel Ojelel Serves as the technical advisor for environmental projects at Save A Seed for the Future (SAFE), Uganda. Also, serves as Assistant Lecturer in the Department of Plant Sciences, Makerere University.

Yaw Osei-Owusu Works as a Director at Conservation Alliance International, Ghana, on projects focused on assessing the impacts of policies on different actors' capacity to access, use, and exchange productive systems and biodiversity.

Raymond Owusu-Achiaw Works as a Programme Officer at Conservation Alliance International, Ghana, on projects focused on enhancing communities' adaptive capacity through improved access to good agricultural practices and agroforestry.

Marivic Pajaro Executive Director at Daluhay Daloy ng Buhay Inc. in the Philippines. She is working on the "Ridge to Reef" approach to socio-ecological networks, para-professional training, and other stakeholder capacity building. Her leadership specialises in harmonised designs for community and sustainable development. PhD from the University of British Columbia.

Dambar Pun CEO of Back to Nature Nepal. Has worked on conservation and ecotourism-based business since 2003. Conducting philanthropic work around Panchase Protected Forest for awareness and skill development of community on biodiversity conservation and ecotourism since 2011.

Dem Bahadur Purja Pun Works at Back of Nature Nepal. Has worked in the tourism field since 2007. Conducts awareness work on ecotourism and biodiversity conservation. Passionate and active on tourism development in Nepal.

Mark Edison R. Raquino Research and Development Coordinator at Daluhay Daloy ng Buhay Inc. in the Philippines. His work has focused on optimising the linkages between research and development. His focus includes harmonising science, culture, and traditions through participatory and inclusive efforts with emphasis on youth engagement. Master's degree from the University of the Philippines.

Hugo Remaury Works as a Regional Technical Advisor at the United Nations Development Programme, USA, on local action initiatives that support actors on three essential solution pathways: empowerment, resilience, and investment.

Pia Sethi Has a doctorate in ecology from the University of Illinois, Chicago. Her research addresses human impacts on forest ecosystems and plant-animal interactions. Her current focus is community conservation, hunting impacts, and traditional ecological knowledge.

Pabin Shrestha Works at the Conservation Development Foundation Nepal. Has worked in the fields of taxonomy, ecology, natural resource management, and environmental conservation. Master's in Botany from Tribhuvan University.

Xin Song Programme Specialist at Farmers' Seed Network, China. Working on projects focused on community-based research to build policy-practice-science linkage and promote socio-ecological circular farming that enhances farmers' market capability and sustainable food systems.

Reymar B. Tercero Agricultural Technician at the Aurora Provincial Agriculture Office. With a background in both agriculture and fisheries, he works with farmers and fisherfolk in building their capacity towards resiliency and the use of social artistry as a development process. He represents his agency in providing livelihood support and capacity building for both upland and coastal communities.

Aashish Tiwari Works at the Conservation Development Foundation Nepal. Enthusiastic on natural resource management based on conservation. Master's degree from Institute of Forestry Pokhara on forest management and biodiversity conservation.

Teodoro G. Torio Coastal Resource Management Officer at Aurora Environment and Natural Resources Office. He has been working in coastal resource management for more than two decades and is an expert on Philippine resource law. His knowledge, experiences, and dedication are foundational to the province in programme implementation and marine conservation.

Prasert Trakansuphakon Director of Pgakenyaw (Karen) Association for Sustainable Development (PASD), Thailand. Of Karen origin, a specialist of Indigenous Studies in Thailand and SE Asia, with expertise in academia and civil society. He holds a doctorate in sociology.

Tamara Tschentscher Works as a Consultant at the United Nations Development Programme, Germany, supporting biodiversity conservation projects as a knowledge management and communications expert.

P. Vipindas Works as a Development Coordinator at MS Swaminathan Research Foundation—Community Agrobiodiversity Centre, India, and handles projects concerned with food and nutrition security and watershed development.

Paul Watts President at Daluhay Daloy ng Buhay Inc. in the Philippines. His career focus is on the institutionalisation of socio-ecological and ethnoecology programmes, through an *ecohealth balance* paradigm. In several countries, he has worked extensively on Indigenous and local collaboration for sustainability and food security. Doctor of Science from the University of Oslo.

Chapter 1
Introduction

Maiko Nishi and Suneetha M. Subramanian

1.1 Global Call for Ecosystem Restoration

Ecosystems have been deteriorating worldwide at an alarming rate and consequently harming not only biodiversity but also human well-being. Land degradation resulting from anthropogenic activities has led to a decline in productivity on 23% of the Earth's terrestrial surface, whereas a significant amount of global crop yield has been threatened by pollinator loss (IPBES 2019). Overall, degradation of the land surface is undermining the well-being of the global population of 3.2 billion, inflicting an economic loss that amounts to more than 10% of the annual global gross product (IPBES 2018). Among the ecosystems that significantly suffer from land degradation (including forest and rangeland), wetlands have been most severely degraded; 87% of wetlands have been lost over the past three centuries, while 35%, where data are available, have been lost since 1970 (IPBES 2018, 2019; Ramsar Convention on Wetlands 2018).

Besides the extensive alteration in land use, a significant area of the sea has also been in decline. In 2014, about 66% of the ocean area was under increasingly adverse cumulative impacts, whereas the ocean free from human pressure accounted for merely 3% of the total (IPBES 2019). Together with coastal ecosystems, marine areas have been exposed to various pressures, such as growing agriculture and aquaculture, urban expansion, plastic pollution, and ocean acidity (UNEP 2010; Waltham and Sheaves 2015). This is epitomised by the fact that 37.8% of mangrove forest extent globally was impacted by human activities over the period 1996–2010

M. Nishi (✉) · S. M. Subramanian
United Nations University Institute for the Advanced Study of Sustainability (UNU-IAS),
Tokyo, Japan
e-mail: nishi@unu.edu

© The Author(s) 2023
M. Nishi, S. M. Subramanian (eds.), *Ecosystem Restoration through Managing Socio-Ecological Production Landscapes and Seascapes (SEPLS)*, Satoyama Initiative Thematic Review, https://doi.org/10.1007/978-981-99-1292-6_1

(Thomas et al. 2017). Decline in these ecosystems has negatively affected the vital ecosystem services, including coastal protection, fisheries production, blue carbon capture, and detoxification (Waltham and Sheaves 2015). In particular, reduced coastal protection has been amplifying climate-related risks (e.g. floods and hurricanes) to life and property for 100–300 million people living on coasts within 100-year flood zones (IPBES 2018).

In response, the United Nations (UN) declared 2021–2030 as the "UN Decade on Ecosystem Restoration", adopting a resolution at the 73rd session of the UN General Assembly held on March 1, 2019. This is a global call for action to support and massively scale up efforts to prevent, halt, and reverse the degradation of ecosystems worldwide. With the understanding that healthy ecosystems are indispensable to realise sustainable development, the declaration aims to facilitate global cooperation for ecosystem restoration in line with the 2030 Agenda for Sustainable Development, and for contributions to enhancing livelihoods and tackling the challenges of climate change and biodiversity loss. Led by the UN Environment Programme (UNEP) and the Food and Agriculture Organization of the UN (FAO), the strategy was developed between 2019 and 2020 based on wide consultations to guide the implementation of the Decade for inclusive, joint coordinated action (UNEP and FAO 2020). Since the official launch in June 2021, partnership and collaboration for the Decade implementation have been evolving, involving over 150 official partner organisations as well as thousands of people who participate in a science-based global movement called #GenerationRestoration (UNEP and FAO 2022). An Action Plan was also published in August 2022 to lay out the next steps for collective action in which all stakeholders are mobilised for cooperation, coordination, and synergies in achieving the goals and visions of the Decade (UNEP and FAO 2022).

In this context, ecosystem restoration is defined as "the process of halting and reversing degradation, resulting in improved ecosystem services, and recovered biodiversity" (UNEP 2021, p. 7). This includes a broad range of activities that serve to protect intact ecosystems and remedy degraded ones (e.g. assisting natural regeneration, increasing fish stocks in overfished areas, controlling invasive species, and green infrastructure) (UNEP and FAO 2022). Importantly, the benefits of ecosystem restoration in this framing are explicitly recognised with respect to not only ecosystem functions per se but also their contributions to people. This shows the shift from the classical conceptualisation of ecosystem restoration that used to primarily hinge on natural science, moving towards a dual conceptualisation involving more social dimensions (e.g. values people attach to nature) beyond science (Martin 2017). This is also expressed in the vision for the Decade: "a world where— for the health and well-being of all life on Earth and that of future generations—the relationship between humans and nature has been restored, where the area of healthy ecosystems is increasing, and where ecosystem loss, fragmentation and degradation has been ended" (UNEP and FAO 2022, p. 5). As such, ecosystem restoration is considered what we need for both people and nature.

Over 115 countries have already committed to restoring one billion hectares of degraded land—an area larger than that of the USA or China—by 2030 (Sewell et al. 2020; UNCCD 2022a). Additionally, the 15th Conference of the Parties (COP15) of the UN Convention to Combat Desertification (UNCCD) held in May 2022 decided to accelerate this commitment in the coming years towards 2030 (UNCCD 2022b). The scientific basis also suggests that one of the key actions towards 2030 to stabilise global warming well below 2 °C is to improve the land management (including avoiding loss and degradation) of about 2.5 billion hectares of forests, farms, wetlands, and grasslands, while restoring at least 230 million hectares of natural cover (Griscom et al. 2019).

Nevertheless, estimating what can be achieved in this Decade is a challenge, considering all the dimensions of ecosystem restoration—including the socio-economic, ecological, and cultural. Furthermore, despite our desire to better understand to what extent restoration efforts would be successful, evidence to properly guide policymakers and practitioners for sound restoration is scarce (Cooke et al. 2019). Lacking monitoring and evaluation in restoration projects would lead to lost opportunity, and in the end vast investments may end up with negligible outcomes (Cooke et al. 2019). It is often the case that projects without explicit objectives and accountability do not yield the expected success for restoration (Young and Schwartz 2019). Several related challenges are to precisely understand the baseline status, extent, and level of ecosystem degradation, define clear and realistic objectives for restoration, find out pathways to achieve them, and identify ways and means to implement restoration strategies (Abhilash 2021; Young and Schwartz 2019).

1.2 Relevance of Socio-Ecological Production Landscapes and Seascapes to Ecosystem Restoration

The concept of socio-ecological production landscapes and seascapes (SEPLS) defines an ideal landscape approach that promotes sustainable use of natural resources in a manner that retains the mosaic nature of the landscape enabling well-functioning ecosystems. Landscape approaches, [1] which are generally also applicable to seascapes, can broadly be described as activities that integrate developmental and conservation priorities at the scale of the landscape, involving a diverse range of stakeholders in an inclusive manner. Juxtaposing this with the

---

[1] Landscape approaches are used as a term to capture integrated action in a landscape or seascape where multiple stakeholders negotiate and collaborate to ensure that the multiple benefits derived from the landscape and/or seascape are obtained with minimal trade-offs and are equitably shared. It is noteworthy that landscape approaches do not relate exclusively to actions on terrestrial ecosystems.

goal of ecosystem restoration essentially implies focusing on preventing degradation by addressing various related triggers, mindful of diverse interests of multiple actors across and beyond the landscape or seascape. Hence, several landscapes and/or seascapes are subject to different degrees of degradation as a result of multiple factors, including land and sea use policies and regulations, introduction and spread of invasive alien species, adaptation to climate change-related challenges, preference of certain types of uses over others by certain actors, and demographic changes that affect cultural affinities and practices related to the area, among others. In this volume, we highlight efforts being undertaken on the ground to ensure a high level of social-ecological resilience in their own contexts—which means the activities being undertaken help in securing ecological functioning and a good quality of life for all stakeholders.

The case studies are based on the experiences of partners of the IPSI network who are committed to fostering integrated landscape approaches that enable securing multiple benefits derived through SEPLS management and encourage multi-stakeholder and multi-sectoral approaches to their management and governance. These case studies offer rich and practical evidence to help guide further restoration efforts, while advancing relevant knowledge (both from indigenous, traditional, and modern sources) and practices. The experiences in managing SEPLS demonstrate how different stakeholders (including underrepresented groups) identify problems, negotiate multiple value perspectives of nature (whether as something that provides for our well-being, or as deserving its own right to exist, or as part of a broader notion of the interconnectedness of people and nature defined through cultural beliefs and practices), and set out goals for concerted efforts in restoration. They also showcase what actions are taken for restoration on a landscape or seascape scale and whether and how the efforts meet the needs and interests of the stakeholders. Such interests include defining what types of benefits are sought from these efforts and how these benefits are equitably distributed among the various actors.

1.3 Objectives and Structure of the Book

This book seeks to highlight how the efforts in managing SEPLS can prevent, halt, and reverse land and sea degradation, contributing to ecosystem restoration and sustainable development. Through the practical experiences described in different case studies (Chaps. 2–13) and an analysis of key messages arising from them (Chap. 14), we seek to provide a span of issues and possible solutions (including challenges to addressing restoration, good practices, key success factors, innovative incentives, and lessons learnt) towards achieving ecosystem restoration goals from the perspective of local implementation.

- The case studies broadly address the following questions:
  - How and why has the restoration effort in their area been initiated or emerged through the management of SEPLS?
  - What multiple values of nature are expressed, and how are they negotiated and embraced by the stakeholders to define the objectives of the restoration efforts in managing SEPLS?
  - How effective is SEPLS management to prevent, halt, and reverse any degraded ecosystems and achieve restoration objectives? Has local and traditional knowledge and cultural diversity helped to inform or facilitate the restoration of degraded SEPLS? If so, how?
  - What methodologies and approaches have been used for restoration, and how can good practices be replicated? How can we measure, monitor, evaluate, and report the progress and outcomes of the restoration efforts in SEPLS management?
  - What are the challenges and opportunities in restoring ecosystems through managing SEPLS to achieve biodiversity conservation and sustainable development?

A summary overview of the case studies is given in Table 1.1, while the map indicates the locations of the case studies (Fig. 1.1).

Table 1.1 Overview of the case studies

Landscape or seascape	Chapter (Country)	Ecosystem types	Problems/challenges	Key restoration approaches	Major outcomes
Landscapes	Chap. 2 (Ghana)	Shrubland, forest, farmland, fallow land, riparian systems	Illegal logging, unregulated charcoal production, bushfires	Awareness raising, biodiversity baseline survey, training (e.g. green business models, fire management, landscape monitoring application tools)	A ban on charcoal production declared by a traditional authority, introduction of improved cookstoves, reduced bushfire frequency with the enhanced community capacity for fire control, extended restoration activities (e.g. natural regeneration, agroforestry, enrichment planting, woodlot establishment)
	Chap. 3 (Kenya)	Rangeland (grasslands, woodlands), forest, cropland	Overgrazing (soil erosion), limited natural resource management (overexploitation of natural resources, logging, charcoal production), climate change (drought), diminishing traditional ecological resource governance structures	Soil and water conservation, technical training and capacity development (e.g. awareness campaigns, sustainable grazing practices), community incentives (e.g. wages for restoration jobs, scholarships), technical and financial support (e.g. gullies healing for soil erosion control)	Land restoration (total of 8328 hectares), endorsement of community conservation agreements, development of restoration plans, recruitment and training of community members, establishment of women-led grass seedbanks, sharing and dissemination of lessons learnt (e.g. publications on monitoring results of carbon gains)
	Chap. 4 (Nepal)	Forest, mountain, agriculture	Forest degradation, invasive species, outmigration	Sustainable ecotourism, peace parks	Establishment of peace biodiversity park with amenities for tourists provided by local communities, reduced siltation, water recharge

	Chap. 5 (Uganda)	Agroforestry parklands	Increase demand for fuelwood and charcoal, demographic changes, loss of traditional knowledge	Documentation of tree species and related traditional knowledge, selection and planting of appropriate species	Diverse tree planting
	Chap. 6 (Cameroon)	Farmland, forest, grassland, mountain	Low soil fertility (continuous tillage, soil erosion, overgrazing, cultivation of marginal lands, unsustainable use of natural resources, deforestation, population growth)	Management practices to improve soil fertility	Adoption of agroforestry practices, organic manure production, nursery establishment
	Chap. 7 (India and Thailand)	Agriculture, forest	Regulatory pressure to reduce fallow periods, conversion to settled agriculture, stereotyping that shifting cultivation is environmentally harmful, demographic changes	Traditional practices to enhance restoration in shorter time frames	Enhanced food and nutritional security through maintenance of soil fertility
	Chap. 8 (India)	Agriculture	Loss of wetland ecosystems suited for rice cultivation, outmigration, reduction in traditional rice varieties and related species	Rice ecosystem-based approach, conservation of traditional rice varieties in Rice seed villages	On-farm conservation among farming communities of traditional and modern variety, seed quality improvement through seed exchange networks, food and nutrition security, climate adaptation
Seascapes	Chap. 9 (Ghana)	Wetland (Ramsar site), Man and the Biosphere reserve	Overfishing, pollution, mangrove exploitation, invasive species and bushfires, industrial activities like mining and tourism	Multi-stakeholder interviews using community ecosystem service values typology, community resource management committees overseeing management of the landscape	Mangrove restoration, alternative fuelwood planting, conservation of marine turtle, clearing of aquatic weeds, location of disoriented turtles, pest control, ecological monitoring

(continued)

Table 1.1 (continued)

Landscape or seascape	Chapter (Country)	Ecosystem types	Problems/challenges	Key restoration approaches	Major outcomes
	Chap. 10 (Kenya)	Coastal, marine ecosystems (coral reefs, mangrove, beaches, estuaries)	Destructive fishing (e.g. overexploitation of some fish species), unsustainable tourism practices, illegal logging, climate change	Promotion of mangrove forest and coral reef rehabilitation, ecotourism enterprise development, value chain development for sustainable fisheries and fish processing, waste management, multi-stakeholder collaboration	Coral reef rehabilitation (e.g. improved natural recovery of reefs), mangrove restoration (50% survival rate, creation of 110 casual jobs for local communities), fish value chain addition (increased incomes for women), improved waste management (recycling, collective cleaning activities), ecotourism development (e.g. boat tours by the locals), enforcement of by-laws in joint fisheries co-management areas (CMAs), participatory governance mechanism
	Chap. 11 (Chinese Taipei)	Coastal, marine	Intensive fishing activities (e.g. via methods to increase catches and improve fishing efficiency), climate and environmental changes	Participatory action research for public–private collaboration (e.g. importance-performance analysis, focus group meetings), environmental education (e.g. awareness raising, ecological survey training), organisation of an ecological survey team	Adoption and expansion of citizen science (e.g. co-creation of panel data on ecosystems for monitoring and further research)

	Chap. 12 (China)	Coastal, marine	Overfishing, pollutants discharge	Creation of a multi-stakeholder platform for conservation, support for formation of a participatory fishery conservation group, participatory monitoring of marine species, waste management training	Establishment of eco-friendly guesthouse, launch of marine-friendly reading room, establishment of eco-garden in an elementary school
Landscapes and Seascapes	Chap. 13 (Philippines)	Forest, mountains and coastal (ridge to reef)	Inadequate expertise among communities to design, plan, and monitor ecosystem stewardship and restoration activities	Capacity development of Indigenous Peoples and Local Communities (IPLCs) on conservation, scaling up, and networking across larger ecosystems and governance areas	Restoration of forests in ancestral areas, enhanced stewardship efforts of IPLCs with one focus on ridge to reef food systems of communities

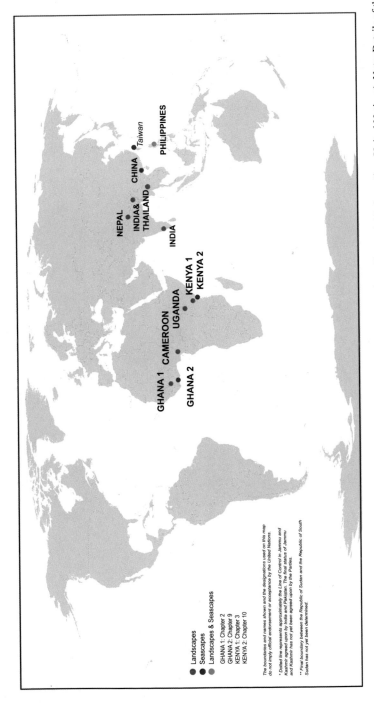

Fig. 1.1 Locations of the case studies (regions, and landscapes and/or seascapes) (Map template: Geospatial Information, United Nations). Note: Details of the case study locations, including geographic coordinates, are described in each chapter

References

Abhilash PC (2021) Restoring the unrestored: strategies for restoring global land during the UN decade on ecosystem restoration (UN-DER). Land 10:201. https://doi.org/10.3390/land10020201

Cooke SJ, Bennett JR, Jones HP (2019) We have a long way to go if we want to realize the promise of the "Decade on Ecosystem Restoration". Conserv Sci Pract 1:e129. https://doi.org/10.1111/csp2.129

Griscom BW, Lomax G, Kroeger T, Fargione JE, Adams J, Almond L, Bossio D, Cook-Patton SC, Ellis PW, Kennedy CM, Kiesecker J (2019) We need both natural and energy solutions to stabilize our climate. Glob Chang Biol 25:1889–1890. https://doi.org/10.1111/gcb.14612

IPBES (2019) Summary for policymakers of the global assessment report on biodiversity and ecosystem services of the intergovernmental science-policy platform on biodiversity and ecosystem services. IPBES secretariat, Bonn. https://doi.org/10.5281/zenodo.3553579

IPBES (2018) Summary for policymakers of the assessment report on land degradation and restoration of the intergovernmental science-policy platform on biodiversity and ecosystem services. IPBES secretariat, Bonn

Martin DM (2017) Ecological restoration should be redefined for the twenty-first century. Restor Ecol 25:668–673. https://doi.org/10.1111/rec.12554

Ramsar Convention on Wetlands (2018) Global Wetland Outlook: State of the World's Wetlands and their Services to People

Sewell A, van der Esc S, Löwenhardt H (2020) Goals and commitments for the restoration decade: a global overview of countries' restoration commitments under the Rio conventions and other pledges. PBL Netherlands Environmental Assessment Agency

Thomas N, Lucas R, Bunting P, Hardy A, Rosenqvist A, Simard M (2017) Distribution and drivers of global mangrove forest change, 1996–2010. PLoS One 12:e0179302. https://doi.org/10.1371/journal.pone.0179302

UNCCD (2022a) Summary for decision makers. Global land outlook, 2nd edn. UNCCD. United Nations Convention to Combat Desertification, Bonn

UNCCD (2022b) United global call to act on land degradation and drought concludes major UN meeting in Côte d'Ivoire [WWW document]. UNCCD, viewed 14 September 2022. Retrieved from https://www.unccd.int/news-stories/press-releases/united-global-call-act-land-degradation-and-drought-concludes-major-un

UNEP (2021) Becoming #GenerationRestoration: ecosystem restoration for people, nature and climate. UNEP - UN Environment Programme, Nairobi

UNEP (2010) NEP emerging issues: environmental consequences of ocean acidification: a threat to food Security. UNEP, Nairobi

UNEP & FAO (2022) Action plan for the UN decade on ecosystem restoration, 2021–2030—Version August 2022

UNEP, & FAO (2020) The United Nations Decade on Ecosystem Restoration: Strategy

Waltham NJ, Sheaves M (2015) Expanding coastal urban and industrial seascape in the great barrier reef world heritage area: critical need for coordinated planning and policy. Mar Policy 57:78–84. https://doi.org/10.1016/j.marpol.2015.03.030

Young TP, Schwartz MW (2019) The decade on ecosystem restoration is an impetus to get it right. Conserv Sci Pract 1:e145. https://doi.org/10.1111/csp2.145

The opinions expressed in this chapter are those of the author(s) and do not necessarily reflect the views of UNU-IAS, its Board of Directors, or the countries they represent.

Open Access This chapter is licenced under the terms of the Creative Commons Attribution-NonCommercial-ShareAlike 3.0 IGO licence (http://creativecommons.org/licenses/by-nc-sa/3.0/igo/), which permits any noncommercial use, sharing, adaptation, distribution and reproduction in any medium or format, as long as you give appropriate credit to UNU-IAS, provide a link to the Creative Commons licence and indicate if changes were made. If you remix, transform, or build upon this book or a part thereof, you must distribute your contributions under the same licence as the original. The use of the UNU-IAS name and logo, shall be subject to a separate written licence agreement between UNU-IAS and the user and is not authorised as part of this CC BY-NC-SA 3.0 IGO licence. Note that the link provided above includes additional terms and conditions of the licence.

The images or other third party material in this chapter are included in the chapter's Creative Commons licence, unless indicated otherwise in a credit line to the material. If material is not included in the chapter's Creative Commons licence and your intended use is not permitted by statutory regulation or exceeds the permitted use, you will need to obtain permission directly from the copyright holder.

Chapter 2
Community-Based Woodland Restoration for Livelihoods and Sustainable Wood Fuel Utilisation in the Mole Ecological Landscape, Ghana

Jacqueline Sapoama Mbawine and Godwin Evenyo Dzekoto

2.1   Introduction

In Ghana, biodiversity conservation is carried out mainly through nationally designated protected areas and the institution of indigenous traditional measures such as the creation of community sacred groves, reverence of totem animals,[1] and the observance of taboo days[2] (Adom 2018). The Community Resource Management Area (CREMA) model has been recognised for its role in management of biodiversity on community lands outside protected areas. Outside traditional conservation management methods, the CREMA framework creates a platform on which stakeholders, including both state and non-state actors, can express and negotiate their interests, such as embracing collective values on ecosystem restoration as outlined in this case study.

This case study is based on outcomes of the Mole Community-Based Woodland Restoration for Sustainable Wood Fuel Utilisation project developed and implemented by A Rocha Ghana in CREMAs within the Mole Ecological Landscape (Fig. 2.1 and Table 2.1).

The Mole Ecological Landscape is a biodiversity hotspot in the Savannah Region of Ghana that supports a myriad of local livelihoods. However, the landscape faces challenges such as illegal logging, unregulated charcoal production, and bushfires. These challenges threaten biodiversity, contribute to climate change, and undermine

---

[1] Totem animals are not hunted due to their cultural significance.

[2] Days on which community members do not go to farm, hunt or fish.

J. S. Mbawine (✉) · G. E. Dzekoto
A Rocha Ghana, Accra, Ghana
e-mail: jacqueline.kumadoh@arocha.org

© The Author(s) 2023
M. Nishi, S. M. Subramanian (eds.), *Ecosystem Restoration through Managing Socio-Ecological Production Landscapes and Seascapes (SEPLS)*, Satoyama Initiative Thematic Review, https://doi.org/10.1007/978-981-99-1292-6_2

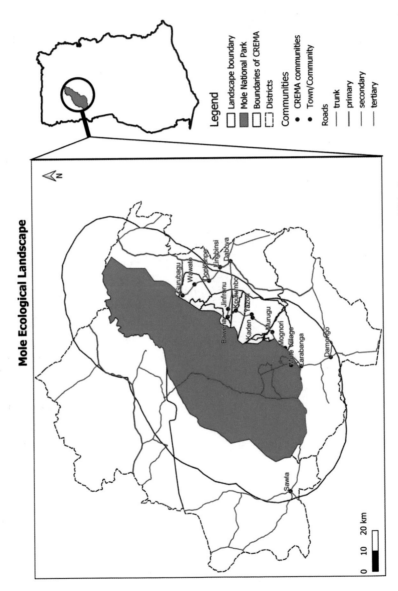

Fig. 2.1 Map of the Mole Ecological Landscape (Source: A Rocha Ghana and Noe 2022, unpublished)

Table 2.1 Basic information of the study area (Source: A Rocha Ghana 2020)

Country	Ghana
Province	Savannah Region
District	N/A
Municipality	West Gonja Municipality
Size of geographical area (hectare)	1,213,826
Dominant ethnicity(ies), if appropriate	Gonja, Hanga
Size of case study/project area (hectare)	122,240
Dominant ethnicity in the project area	Gonja
Number of direct beneficiaries (Selected Community members)	1,000
Number of indirect beneficiaries (Wider population of the five communities where project was implemented)	20,000
Geographic coordinates (latitude, longitude)	9° 42′ 0″ N, 1° 50′ 0″ W

the sustainability of community livelihoods. These threats were further intensified between 2003 and 2021 during a spate of increased illegal logging of rosewood, an endangered species listed under Convention on International Trade in Endangered Species of Wild Fauna and Flora (CITES) Appendix II (Obiri et al. 2022; Adjonou et al. 2020; Dogbevi 2020). The project was developed with the aim of engendering sustainable vegetation harvesting practices and sustainable regimes for wood fuel and charcoal production within the Mole Ecological Landscape as a means of addressing the identified landscape challenges. Specific objectives of the project that are relevant to this chapter are to: (1) create awareness and build capacity for the restoration of 500 ha of woodlands using agroforestry, enrichment planting, and assisted natural regeneration approaches; (2) mobilise communities for participatory mapping and planning of restoration activities; and (3) organise and train charcoal producers for a more sustainable charcoal production value chain in the Mole landscape.

Several efforts have been made by the government and traditional authorities to address the landscape challenges, including national bans on rosewood logging (Obiri et al. 2022), traditional bans on commercial charcoal production (Fugu 2018; Adams 2021), and closure of commercial businesses in charcoal and rosewood harvesting (Fugu 2021). However, similar efforts have not been made to restore the degraded landscape of the Mole Ecological Landscape prior to this project (Dogbevi 2020). Although the project period is 2019–2023, the interventions outlined in this case study cover the period of 2019–2021. These interventions are essential contributions to addressing the landscape challenges for maintaining not only the remnant biodiversity, but also conserving the landscape characteristics. The project was funded by the Federal Ministry for the Environment, and Nature Conservation and Nuclear Safety (BMUV) of Germany through the German Development Cooperation (GIZ).

2.2 Methods

2.2.1 Case Study Area Description

The Mole Ecological Landscape in the Guinean savannah ecozone is one of the critical biodiversity hotspots in Ghana. The landscape hosts Ghana's first, largest, and most prestigious protected area, the Mole National Park (Mole National Park 2015). Conservation of biodiversity within the landscape is largely through this nationally designated protected area, which covers approximately 4577 km². The park is home to 742 vascular plant species, over 90 mammal species, including five primates, 334 bird species, 33 reptiles, 9 amphibian species, and about 120 butterfly species (Mole National Park 2015). Within the landscape, some wildlife species of conservation concern in the fringe communities include: the caracal (*Caracal caracal*), common genet (*Genetta genetta*), bohor reedbuck (*Redunca redunca*), roan antelope (*Hippotragus equinus*), Nile crocodile (*Crocodylus niloticus*), and Nile monitor lizard (*Varanus niloticus*). Besides the wildlife, tree species of conservation concern include the African mahogany (*Afzelia africana*) and rosewood (*Pterocarpus erinaceus*). Economically important trees include the shea tree (*Vitellaria paradoxa*) and dawadawa (*Parkia biglobosa*).

The Mole Ecological Landscape has two main seasons, the rainy season (April to October) peaking in June or July with dry spells in August, and the dry season (November to March) (MOFA 2022). The landscape provides a wide range of ecosystem services for communities; key amongst them are agricultural production, ecotourism, and mitigation of climate change impacts. The landscape also supports the livelihoods of over 80,000 community members who are mainly farmers, charcoal producers, beekeepers, or pastoralists (A Rocha Ghana 2020). Another significant livelihood activity within the landscape is the collection and processing of *Vitellaria paradoxa* (sheanut) into shea butter by the women.

The CREMA model, which was developed by the Wildlife Division of the Forestry Commission of Ghana as a policy intervention for conservation challenges, is aimed at promoting collaborative and participatory wildlife management within communities fringing protected areas in the country (Asare et al. 2013; Forestry Commission 2004). The model, which works as a community-based organisation, principally involves a community or group of communities agreeing on the management of a common resource area. As part of the management system, the CREMA is organised at three levels: the advisory level (CREMA Executive Committee); community level (Community Resource Management Committee-CRMC); and the individual community member or landholder level (Forestry Commission

2004). The development of a constitution and the establishment of relevant by-laws also guide and regulate natural resource governance and management activities in the respective constituent communities (MLNR 2019). Communities also agree to define CREMA boundaries and demarcate their land into three main zones: the core zone, which is biodiversity-rich area for conservation purposes; the buffer zone, where there is limited resource extraction; and the transition zone, where farming and other resource utilisation is allowed (Forestry Commission 2004).

2.2.2   Interventions in Response to Landscape Challenges

Data for the case study were collected through the various activities implemented by the project to address the identified landscape challenges of illegal logging, unregulated charcoal production, and bushfires. For the purpose of this case study, the activities reflect thematic areas in response to the identified landscape challenges. These thematic areas are landscape restoration in response to illegal logging and other related challenges through biodiversity baseline surveys, awareness creation, community mapping, and restoration activities. Promoting sustainable biomass energy production through training of charcoal producers in green business models was the response to unregulated charcoal production. Bushfire control through fire management training was the response to bushfires. Likewise, promotion of citizen science landscape monitoring was a crosscutting response to monitor and assess the effectiveness of project interventions in contributing to the sustainability of the landscape.

The baseline surveys used transect walks to collect fauna data by recording sightings and animal signs along the transect. Vegetation data were collected through sample plots by assessing floral diversity, presence of rare, threatened, or endangered species, and assessment of natural regeneration species through growth performance. Community meetings were held to raise awareness on landscape biodiversity and its link to sustainable livelihoods. The meetings were also used to discuss the essence of landscape restoration. Community maps were developed using participatory mapping to identify areas that needed restoration. For the purpose of ensuring that restoration efforts did not have a negative net impact on the landscape and native species, the landscape was stratified into four categories based on biodiversity and restoration needs. Appropriate restoration techniques were then applied to provide the best results and a balanced net impact of the restoration efforts on landscape biodiversity. Details of the activities are outlined below.

Biodiversity Baseline Surveys   As part of efforts to address the identified landscape challenges, the project sought to establish baselines in the CREMA. These were to serve as the basis for preparation of stakeholder engagement and development of strategies for undertaking restoration activities, which are one of the key focuses of the project. The baseline survey, which covered both flora and fauna, was carried out in August 2020 in five communities participating in the restoration activities. These communities were strategically selected based on their impact on the landscape as a result of resource utilisation. The CREMA was stratified into four main land use categories, namely farmlands, fallow lands, riparian areas, and core zone. A stratified sampling approach was employed to best select survey sites and to characterise species occurrence and distribution within the four land uses. The fauna survey was carried out by walking a 1–2 km transect per observation period, taking note of species encountered on a data sheet and recording their Global Positioning System (GPS) waypoints. For the vegetation survey, a plot-based vegetation sampling approach was used to sample vegetation communities, floral diversity, and the possible presence of rare, threatened, or endangered species. A GPS was used to record all plot locations and boundaries. The survey also estimated tree density and composition of different species present in the identified land use categories. All-natural regenerating species were identified and counted, and their growth performance was assessed. Seedlings and saplings ≥ 0.10–1.3 m in height inside the subplots were tallied as regeneration.

Awareness Creation Awareness creation and community engagement was carried out by building onto the existing CREMA structure. Awareness creation activities employed methods such as community durbars and village square meetings on pre-agreed scheduled dates. At the meetings, engagement methods included the use of documentaries to showcase conservation advocacy, the importance of conservation, and National Geographic videos, amongst others. These were a prelude to sharing the results of the baseline survey, which served as an entry point to engage community members on the need for restoration activities, as well as the project and its activities. Opportunities for clarifications, questions, and answers were provided as part of getting community members to buy into the project's proposed actions to address the identified challenges. Community engagement was also used as an opportunity to solicit volunteers who would be willing to participate in implementation of the project interventions, particularly for the restoration activities.

Community Land Use Mapping The project carried out a landscape restoration needs assessment in October 2020 and supported the communities to develop a land use map based on landscape characteristics and land use systems. The aim of the community land use mapping was to identify opportunities for restoration activities. This was a participatory process undertaken first with communities in a community meeting setting, and validated in a workshop setting with the involvement of state and non-state actors and beneficiary communities. Another focus of this activity was to stimulate discussions around sustainable wood fuel harvest and charcoal production.

Restoration Activities The restoration activities were based on the results of the landscape restoration needs assessment and the community land use mapping exercises. A community nursery was established to be managed by women to provide a continuous supply of seedlings for the restoration efforts (Fig. 2.2).

Fig. 2.2 Women carrying seedlings for restoration (Source: A Rocha Ghana 2020, Photo credit: IUCN)

The project employed restoration techniques such as the Farmer Managed Natural Regeneration (FMNR),[3] agroforestry, enrichment planting,[4] and woodlot establishment. The techniques employed in the restoration activities were aimed at maximising restoration efforts and achieving a net balance of landscape restoration. FMNR was employed in degraded areas. Agroforestry was employed on farms, whilst enrichment planting was carried out in very degraded areas. Woodlot establishment was done in areas where fuelwood was sourced. These were mostly fallow lands that also served as greenbelts to protect fire-prone areas within the community. Species planted in woodlots have highly fire-resistant characteristics. Native tree species with both biodiversity benefits (African mahogany and rosewood) and economic benefits (shea tree and dawadawa) were used in the restoration. A total of 500,000 native seedlings were planted as part of restoration activities. An area of 650 ha of land, comprising farmlands, fallow areas, and riverine areas, was planted between 2020 and 2021.

Training for Charcoal Producers Training on green business models was carried out for 11 charcoal producers in the landscape. The training was aimed at promoting

---

[3]Farmer Managed Natural Regeneration (FMNR) comprises a set of practices (such as selective pruning of stems) used by farmers to encourage the growth of native trees on agricultural land. http://fmnrhub.com.au/wp-content/uploads/2013/09/Rinaudo-2007-Development-of-FMNR.pdf.

[4]Enrichment planting is the practice of planting trees with the aim of increasing the density of desired tree species in the area of interest (Tigabu et al. 2010).

the adoption of sustainable energy sources in the landscape. The green business model, which was adapted from the Sustainable Business Model[5] (Breuer and Lüdeke-Freund 2014), supports the development of an idea into a viable business model. It follows a holistic approach regarding the relationships within and outside the business. Besides economic criteria, the model also focuses on the ecological and social consequences of an activity. The business model canvas template (Osterwalder 2013) was used to guide participants on how to develop investment worthy proposals for funding. The template highlighted the importance of identifying key partners required to make the business sustainable, key activities, and resources required to set up and run a business.

Fire Management In collaboration with the District Ghana Fire Services Department, a total of 90 community volunteers were trained from six communities. The volunteers were trained in fire management techniques such as creation of fire belts and reading and interpretation of the Fire Danger Index (FDI). The FDI is calculated using the Fire Burning Index (FBI) plus the Wind Correction Factor (WCF) (i.e. FDI=FBI + WCF) based on information from a local weather station regarding wind, temperature, and relative humidity. If there is rainfall, it becomes Final Fire Danger Index (FFDI). Hence, FFDI = FBI + WCF × RCF, where RCF is the Rain Correction Factor. The FDI chart is colour-rated with values.[6] Once the FDI value for the day is obtained, it is matched with the corresponding colour rating and then interpreted to show the potential for fires on that day. Volunteers were provided with firefighting equipment, such as fire extinguishers, fire beaters, knapsacks for water supply, high visibility uniforms, and Wellington boots, to enable them to combat any wildfires that the communities may experience. Other logistics assistance provided included bamboo bicycles as means of transport, whistles for sending alerts, and cutlasses for creation of fire belts. Monthly allowances to serve as rations are given to the volunteers to support their monitoring activities. Health insurance is also provided annually.

Landscape Monitoring To ensure that efforts to restore the landscape and conserve biodiversity are sustained, the capacity of 35 community volunteers was also built in the use of landscape monitoring application tools, notably the Open Data Kit (ODK) Collect and Event Book, both of which support citizen science data collection. The ODK Collect is an android application that allows for off-line data collection. The application supports audio, images, video, barcodes, signatures, multiple-choice, free text, and numeric answers. The Event Book tool provides communities with flexibility in monitoring biodiversity and record keeping, so that analysis can be carried out and fed into decision-making (Stuart-Hill et al. 2005).

---

[5] A sustainable business model can be defined as a business model that creates, delivers, and captures value for all its stakeholders without depleting the natural, economic, and social capital it relies on.

[6] https://research.csiro.au/bushfire/assessing-bushfire-hazards/hazard-identification/fire-danger-index/

2.3 Results and Discussion

2.3.1 Summary Results—Interventions in Response to Landscape Challenges

The results of the activities are summarised in Table 2.2. The summary gives an overview of the landscape challenges, thematic areas covered by the interventions, the activities implemented to address the identified challenges, stakeholders that were engaged during the implementation, and the outcomes of the actions. The results shared in this summary cover the project implementation so far from 2019 to 2021.

2.3.2 Results from the Interventions and their Implications

Biodiversity Baseline Surveys The biodiversity baseline surveys recorded 20 species of mammals, 70 species of woody plants from 35 families, and 3 species of reptiles. For the fauna survey, mammal species most frequently encountered were rodents: the striped ground squirrel (*Xerus erythropus*) and African giant rat (*Cricetomys emini*), accounting for about 30% and 9% of species recorded, respectively. Other species encountered include ungulates: bushbuck (*Tragelaphus scriptus*) accounting for 25%, and small mammals such as the African savanna hare (*Lepus microtis*), accounting for 8%. Although most of the species recorded are of least concern globally, two species of carnivores (caracal and common genet) and two species of ungulates (bohor reedbuck and roan antelope) are of national conservation concern in Ghana. They are therefore wholly protected in Ghana and thus listed in Schedule I[7] of the Wildlife Conservation Regulations 1971 of Ghana. The caracal is also listed as a CITES Appendix I[8] species, prohibiting its trade (CITES 2022).

The flora survey recorded 70 species of woody plants from 35 families. Shea tree (*Vitellaria paradoxa*), tepiala[9] *(Terminalia avicinoides),* African mahogany (*Afzelia africana),* African copaiba balsam tree *(Daniella oliveri),* and wild syringa *(Burkea africana)* were the predominant species in the landscape. The most abundant woody species enumerated in all plots in the landscape were *Terminalia avicinoides, Daniella oliveri, Pterocarpus erinaceus, Afzelia Africana, Lannea acida,* and *Combretum.* Five of the recorded plant species are of global conservation concern according to the International Union for Conservation of Nature (IUCN, 2020) Red

---

[7] Schedule I: Animals Wholly Protected - No person shall at any time hunt, capture or destroy any of the species mentioned in the First Schedule to these Regulations.

[8] Appendix I includes species threatened with extinction.

[9] Name in Gonja (Burkill 1985).

Table 2.2 Summary of activities undertaken to address landscape challenges (Source: Authors, based on this study)

Landscape challenges	Thematic intervention	Activities implemented	Stakeholders involved	Results
Illegal logging	Landscape restoration	Baseline surveys (flora and fauna)	CREMA monitoring team members	20 species of mammals, 70 species of woody plants from 35 families, and 3 species of reptiles
		Awareness creation	Community members	20,000 community members gained knowledge on impacts of illegal logging and supporting actions to curb it
		Community land use mapping through community meetings, and stakeholder validation workshop	Communities, state actors (municipal assembly, land use planning department), and non-state actors (environment-related businesses within the landscape, e.g. charcoal producers)	Community land use map with areas identified for restoration activities
		Restoration	Community members	500,000 native seedlings planted over 650 ha of land with 60% survival rate
Unregulated charcoal production	Sustainable biomass energy production	Training of 11 charcoal producers on green business model and its application	Charcoal producers	10 business proposals drafted, introduction of improved cookstoves, and promotion of briquette charcoal as alternative to wood fuel
Bushfires	Bushfire control	Fire management training for 90 community members from 6 communities, providing them with allowances and equipment for fire detection and fire management	Community members and district fire services department	Community fire notice board in 6 communities, incentivised volunteers and reduced wildfire incidents

(continued)

Table 2.2 (continued)

Landscape challenges	Thematic intervention	Activities implemented	Stakeholders involved	Results
Landscape monitoring	Citizen science	Landscape monitoring using citizen science data collection methods, notably Open Data Kit (ODK) Collect and Event Book	Community members	Citizen science monitoring data on biodiversity and landscape restoration activities

List categories. These are *Pterocarpus erinaceus* (EN), *Afzelia africana* (VU), *Khaya senegalensis* (VU), *Vitellaria paradoxa* (VU), and *Raphia sudanica* (DD).

The baseline assessment and its accompanying field investigations have confirmed the existence of most of the common species in the savannah ecozone of Ghana, together with species of national and global conservation concern (A Rocha Ghana 2020, unpublished). The presence of keystone flora species such as rosewood and African mahogany, which have been targets of illegal logging (Obiri et al. 2022), indicated an opportunity for their inclusion in the restoration activities to secure them from local extinction. Furthermore, use of the baseline information as part of the restoration strategy served as a basis for monitoring restoration efforts and assessing if any of the interventions had had any undesired impact on the wider landscape beyond the desired net impact of improved vegetation, sustained biodiversity, and landscape restoration (Bull et al. 2013). The evidence from the baseline survey also served as an entry point for awareness raising, thereby facilitating the acceptance of restoration activities by the local communities as a conservation measure to secure biodiversity from further decline.

Awareness Creation The interventions in community engagement and awareness creation involved 20,000 community members who were informed on the impacts of illegal logging on the sustainability of ecosystem services currently provided by the landscape. Awareness creation in landscape restoration has been identified as a key strategy for overcoming engagement challenges regarding socio-cultural norms that lead to landscape degradation and increase the impacts of climate change (Schweizer et al. 2021; Stanturf 2021). The use of awareness creation by the project, therefore, led to the active participation of community members in other interventions introduced by the project, such as restoration activities, particularly woodlot establishment. Active participation is evident in the communities providing land for communal restoration and individuals collecting tree seedlings to be planted on their farms. It also informed the organisation of women for the establishment and management of a community nursery to supply the seedlings needed to undertake the restoration activities. These actions by community members within the study area can also be assessed as a form of adaptive management, which is useful in ensuring the sustainability of landscape restoration efforts (Williams and Brown 2014).

Community Land Use Mapping The landscape needs assessment and community mapping carried out by the project led to the identification of areas categorised as farmlands, degraded areas, very degraded areas, and fallow lands. Farmlands were defined as areas suitable for farming. Degraded areas were defined as areas in the communities where biodiversity stock was mainly saplings and had lots of under-growth. Very degraded areas were defined as areas where there was a low density of native flagship species. Fallow areas were defined as either fire-prone areas or areas where fuelwood was sourced. The community land use mapping resulted in the development of a community land use map to guide resource utilisation within the landscape. Land use mapping is a popular tool for maximising land use benefits and guiding decision-making regarding land resources (Vihervaara et al. 2018; Poccard-Chapuis et al. 2021). The development of a community land use map using partic-ipatory mapping tools has been identified as essential for assisting stakeholders in valuing landscape elements based on their needs, developing a vision for the landscape, as well as guiding landscape investment priorities (Ernoul et al. 2018). The community land use map is therefore used by communities to identify and monitor changes in land use and land cover that occur within the landscape. The map also guides communities' decisions on identifying areas that need interventions, such as restoration, interventions for climate smart agriculture, biodiversity moni-toring, and monitoring for bushfires. This intervention thereby feeds into the achievement of the landscape vision of *a resilient landscape supporting livelihoods and biodiversity* (A Rocha Ghana and IUCN 2019, unpublished).

Restoration The project target was 500 ha, however, due to the overwhelming acceptance of the project and its interventions by communities, the project exceeded its original target by 150 ha. Beneficiary communities, as part of the restoration activities, planted 500,000 native seedlings on a total of 650 ha of land with a survival rate of 60%. The woodlots were established to provide sustainable raw materials for charcoal producers. These efforts contribute to the achievement of the targets set by the National Strategic Energy Plan (NSEP) to establish over 6.5 million ha of plantations by 2020 (Energy Commission 2006). The NSEP is a government initiative started in 2006 that aims to contribute to the development of a sound energy market that would provide sufficient, viable, and efficient energy services for Ghana's economic development. Some of NSEP's major activities include setting up plantations and the creation of woodlots, as wood fuels account for more than 60% of total energy used in Ghana (Energy Commission 2006).

Effective landscape restoration contributes to enhancing biodiversity and the provision of ecosystem services as described in the examples of Gann et al. (2019). The assessed net effects so far of the restoration efforts of this case study based on the different restoration techniques show a balance between the different areas planted and a general increase in vegetation cover within the landscape (Gann et al. 2019). Other benefits within the local communities include supporting healthy wildlife populations through securing habitats, and maintaining the primary and secondary seed dispersal essential for plant regeneration and woodland dynamics (Ruxton and Schaefer 2012). This goes to confirm the assertion by Stanturf et al.

(2019) that there is no one-size-fits-all approach to landscape restoration, however, the complexity of socio-ecological systems provides not only challenges but also opportunities for meeting diverse needs. Furthermore, the loss of keystone mega-fauna from ecosystems in the landscape could have profound and long-lasting negative effects on its ecological structure and function (Malhi et al., 2016). This chapter argues that based on the results of the project so far, the identified landscape threats, if not mitigated, would greatly affect the landscape's ability to support itself through pollinator activities and tree dispersal to regenerate and serve as a habitat for wildlife (Newton et al. 2018). It is, however, fair to say that the project through the strategies employed, has applied some principles of ecological restoration as outlined by Gann et al. (2019), thereby resulting in the current observed net gain. Key highlights of these principles include the *effective engagement of a wide range of stakeholders* and *the utilisation of scientific, traditional, and local knowledge* (Gann et al. 2019). The restoration efforts of the project are also contributing to the achievement of SDG 15.

Training for Charcoal Producers The green business model training for charcoal producers within the landscape focused on using sustainable means of producing biomass energy and finding alternatives to the use of wood fuel. The training led to the development of alternative biomass energy sources such as briquettes and the development of business plans targeted at investors within the landscape to support the production of briquette instead of fuelwood. Another intervention, the introduction of fuel-efficient cookstoves, involved the distribution of 300 cookstoves as part of a pilot to 300 households. The promotion of alternatives to fuelwood, such as briquettes, as a strategy for landscape restoration is one that has been successfully used in other areas. For instance, South Korea restored degraded landscapes using interventions such as substituting wood with coal briquettes, remodelling house heating systems to use the briquettes, and establishing fuelwood plantations, also known as woodlots (Stanturf et al. 2019). The project's introduction of briquettes and fuel-efficient stoves as alternatives to high dependence on fuelwood is validated by the South Korean example. Furthermore, the creation of sustainable livelihood opportunities through inclusive decision-making is highly recommended as a strategy for sustainable landscape restoration efforts (ITTO 2021). Taking the interventions on introduction of alternative livelihoods further, 11 businesses that took part in the training have since developed proposals suitable for various investment opportunities. One of these 11 businesses has actually received investment support to translate its business idea into a product that could be adopted by the communities to address their biomass energy dependency. The provision of capacity building for these businesses through the landscape restoration initiative is contributing to new income sources, thereby building community resilience to the impacts of global challenges such as climate change (Agostini and Proskuryakova 2022).

Fire Management As a result of the fire management training carried out in 2020, communities were better prepared for fires in 2021. With the creation of fire belts in fire-prone areas and community members' ability to read and interpret the Fire Danger Index, communities were able to successfully combat seven out of eight

Fig. 2.3 Creation of fire belt by Community Fire Volunteers (Source: A Rocha Ghana, 2021, Photo by A Rocha Ghana)

fire incidents that were recorded in 2021. One of the six communities that were trained in fire management did not record any fires at all. Fire as a tool for landscape management has benefits of enhancing landscape vegetation, improving soil conditions, and even promoting the regeneration of some fire-dependent tree species such as *Pterocarpus erinaceus, Combretum nigricans*, and *Terminalia Avicennioides* (Obiri et al. 2022; Husseini et al. 2020; Amoako et al. 2018; Alagona et al. 2012). However, the devastating nature of wildfires in particular cannot be underestimated as they also contribute to land use changes and impact livelihoods. Therefore, the use of fire management is required to address the changes resulting from bushfires (Yaro and Tsikata 2013; Butsic et al. 2015). Studies have shown that the effect of bushfires on vegetation cover loss and the attendant impact on biodiversity loss/climate change is an annual occurrence, particularly in the dry season between November and April (Kugbe et al. 2012). Within the Savanna ecozone where the study area is located, the rate of bushfire incidences has been the highest compared to other parts of Ghana. Bushfires have led not only to loss of biodiversity and vegetation cover, but also life and property, particularly farmlands (Amoako et al. 2018). The essence of the fire management training was therefore to build community resilience and enhance response time to addressing fire incidences through fire management techniques (Fig. 2.3).

To sustain the efforts resulting from the fire management training, a community Fire Danger Index notice board was mounted at the community square, and daily rankings are displayed to create awareness of potential fires. The notice board is updated daily by selected community fire volunteers to inform communities of the potential for bushfires each day. The project also provided fire management logistics and organised community volunteers who take turns monitoring for potential fires, thus resulting in faster response time for combating wildfires. The cumulative impact of these interventions has been an observed reduction in bushfires due to community capacity to read and interpret the fire index and more importantly to respond to fire incidences.

Promotion of Citizen Science Landscape Monitoring The use of the ODK Collect application and the Event Book tool is helping community members to track the types of species that are spilling over into their communities that can be used in marketing to tourists. Given that ecotourism is one of the livelihood activities that the local communities are developing, wildlife spillovers from the Mole National Park that are picked up by the monitoring tools thereby inform on the availability of species to attract and sustain visitor presence at the community level. Furthermore, monitoring results have also shown evidence of good natural regeneration of most woody species enumerated in the landscape based on the fact that 48 tree species out of the 70 recorded are naturally regenerating. This observation is validated by Stanturf et al. (2019), who indicated that monitoring restored systems and allowing time for them to develop were necessary steps to ensure successful restoration.

The use of citizen science to help monitor conservation efforts has evolved over the last three decades (Catlin-Groves 2012). With the introduction of technology, data collection through the use of smartphones has made data transmission a lot easier (Catlin-Groves 2012). Moreover, citizen science enables the collection of data or monitoring of conservation efforts over a large area within the shortest possible time due to a large number of people providing information and options to validate information through the use of GPS (Catlin-Groves 2012). Through the evolution of citizen science, different tools have been introduced, making data collection and transmission a walk in the park. It has been argued that, although citizen science data collection may not be carried out by experts, it has informed decision-making on several conservation actions (Hovis et al. 2020; Catlin-Groves 2012). The use of the ODK Collect application and the Event Book tool has enabled the communities to monitor and provide updates on biodiversity issues so that these can be incorporated into decision-making and the development of landscape management plans.

2.4 Lessons Learnt and Their Applications

Some of the lessons learnt during project implementation include the value of the CREMA as a tool for stakeholder engagement. CREMAs provide a neutral platform where state and non-state actors can discuss landscape management issues with communities and find practical solutions. In the broader sense, the CREMA can be used in a participatory way to achieve both biodiversity and ecosystem conservation goals, as well as to enhance community livelihoods.

Community responses to promote conservation can yield better results if efforts are evidence-based and conducted in a participatory manner with all key stakeholders on board. This is demonstrated in the use of the baseline studies and engagement with stakeholders through awareness creation activities and trainings, which led to traditional leaders placing a ban on charcoal production within the Mole Ecological Landscape. Furthermore, the involvement of the CREMA monitoring team members in data collection served as a good entry point for engaging communities on adopting alternative actions, such as the adoption of improved cookstoves, establishment of woodlots, and restoration activities in general. The evidence from the baseline studies also facilitated dialogue between stakeholders, notably charcoal producers, local government authorities (municipal assemblies) and communities, prompting them to move towards the development of alternatives to charcoal produced from trees—i.e. greener energy sources such as briquette from biomass waste.

The engagement and capacity building of businesses on green models in 2021 also ignited business innovations in seeking greener and more sustainable alternative options to charcoal and wood fuel in the landscape. So far, nine businesses are being supported in obtaining investments for their innovative ideas. More important to the engagement and capacity building has been the creation of investment opportunities for these businesses to ensure that their innovations come alive and have concrete impacts on local communities.

Supporting implementation of the practical actions taught during training sessions served as a learning tool for the wider community. This is notable in the building of the community Fire Danger Index notice board that serves as a continuous guide to community members in their efforts to address bushfires. Provision of monetary allowances and equipment motivated volunteers to undertake regular monitoring exercises, thereby controlling and preventing many fire incidents.

Another important lesson is that to incentivise community members in the restoration of landscapes, focus should not only be on biodiversity and ecosystem conservation, but also on the sustainability of community livelihoods. This is evident in the identification of economically important tree species, such as the shea tree (*Vitellaria paradoxa*) and dawadawa (*Parkia biglobosa*), during the baseline study, which were subsequently incorporated as trees of choice in the restoration. The use of these economically important trees whipped up communities' enthusiasm to restore their landscape for the added benefit of harvesting the fruits of these trees to support their income levels. Also, early planting of seedlings enhanced their establishment, and early creation of a fire belt at restored sites helped prevent fires.

2.5 Conclusions and Recommendations

The project's aim was to engender sustainable vegetation harvesting practices and sustainable regimes for wood fuel and charcoal production within the Mole Ecological Landscape as a means of addressing landscape challenges, specifically through awareness raising, capacity building, and restoration of degraded areas. Within the period targeted by this case study (2019–2021), the project was able to achieve its objectives and exceed some of its targets. Communities under their own leadership took steps such as the ban on charcoal production, the adoption of improved cookstoves, and establishment of woodlots. Their general enthusiasm in undertaking restoration activities allowed for the exceeding of the project's 500 ha target and the achievement of 650 ha of restored land.

The results of the interventions outlined in this chapter were achieved through the involvement of stakeholders, particularly community people who are at the core of the challenges faced within the landscape and are thus most affected by the impacts of these identified challenges. The results of the baseline studies also confirmed that the CREMA assessment areas in the Mole Ecological Landscape are still uniquely important sites as they harbour a number of threatened species, such as *Pterocarpus erinaceus*, within a severely impacted woodland landscape and diminishing habitats.

The commitment demonstrated by community members and stakeholders within the landscape during the implementation of the project resulted from the evidence-based awareness creation and the participatory nature of the project involving all relevant stakeholders. This created an avenue of trust between stakeholders and the project implementers, building community confidence that the interventions introduced were sustainable.

The integrated approach to landscape restoration adopted is contributing to securing landscape vegetation, reducing degradation, enhancing biodiversity and ecosystem conservation, and supporting the livelihoods of fragile communities within the Mole landscape. Through the intervention, state and non-state actors and beneficiary communities are now equipped to plan, map, and monitor community land use systems, as well as undertake restoration activities in a participatory way.

Approaches to curbing the annual wildfire menace in the landscape should be multifaceted. Strategies to be implemented should include wildfire awareness campaigns, capacity building to detect bushfire potential, ground patrols, fire suppression, and enforcement of legislation relating to woodland fires. These will be key to ensuring the success of restoration and natural regeneration interventions in the landscape. Furthermore, the success of restoration activities can be better achieved if community members are incentivised with species that focus not only on biodiversity conservation but also on economically important species that sustain their livelihoods.

Despite the positive gains made by the project in achieving the desired results, some challenges faced during project implementation include: sustaining remuneration or allowances for fire volunteers and monitoring units; wide-spread fire incidents in other parts of the landscape with spilling over effects on intervention areas; and insufficient collaboration between enforcement agencies to combat and enforce the ban on commercial charcoal production.

Notwithstanding the challenges the project has faced so far in its implementation, efforts are being made to promote further engagement with stakeholders, particularly communities outside of the case study area, to address these challenges. It is anticipated that with the continuous monitoring and engagement of community members and other stakeholders, there will be a wider adoption of briquette charcoal made from agricultural/biomass waste within the landscape by December 2022. It is further anticipated that more degraded areas will be planted to increase the restored area from 650 ha to 750 ha, and that fire incidents outside the case study communities will be reduced through the ripple effect as the project's benefits impact other communities.

Acknowledgements The project acknowledges the support from the Federal Ministry for the Environment, Nature Conservation and Nuclear Safety (BMUV) through the German Development Cooperation (GIZ), and the Dutch Ministry of Foreign Affairs through the International Union for the Conservation of Nature (IUCN) Netherlands Committee for providing the funding for the implementation of the project. We extend our thanks to our local partners, Tropenbos Ghana, and the Ghana office of the IUCN. Our gratitude also goes to the communities who took part in the study and the implementation of the project, and the CREMA executives for mobilising community members within the landscape. We thank the Community Resource Monitoring Persons and Fire Volunteers who worked tirelessly as field assistants, businesses and charcoal producers within the landscape who are adopting and implementing sustainable alternatives to wood fuel, the traditional authorities and families who provided land for restoration, the Municipal Fire Service department who provided training to the fire volunteers, the Municipal Assembly for their support on enforcing by-laws, and the staff and management of A Rocha Ghana whose continuous support to the communities is sustaining the project interventions. All their support and work is duly acknowledged.

References

Adams CN (2021) Savannah regional house of chiefs has bans charcoal burning, fuel-wood, rosewood harvesting, viewed 11 October 2022. Retrieved from https://www.ghanaiantimes. com.gh/savannah-regional-house-of-chiefs-has-bans-charcoal-burning-fuel-wood-rosewood-harvesting/

Adom D (2018) Traditional cosmology and nature conservation at the Bomfobiri wildlife sanctuary of Ghana. Nat Conserv Res 3(1):35–57

Adjonou K, Abotsi KE, Segla KN, Rabiou H, Houetchegnon T, Sourou KNB, Johnson BN, Nougbode Ouinsavi CAI, Kokutse AD, Mahamane A, Kokou K (2020) Vulnerability of African rosewood (*Pterocarpus erinaceus*, Fabaceae) natural stands to climate change and implications for silviculture in West Africa. Heliyon 6(6):e04031

Agostini P, Proskuryakova T (2022) Rethinking landscape restoration in Central Asia to improve lives and livelihoods, viewed 4 October 2022. Retrieved from https://blogs.worldbank.org/europeandcentralasia/rethinking-landscape-restoration-central-asia-improve-lives-and-livelihoods

Alagona PS, Sandlos J, Wiersma YF (2012) Past imperfect: using historical ecology and baseline data for conservation and restoration projects in North America. Environ Philos 9(1):49–70

Amoako EE, Misana S, Kranjac-Berisavljevic G, Zizinga A, Ballu Duwieja A (2018) Effect of the seasonal burning on tree species in the Guinea savanna woodland, Ghana: implications for climate change mitigation. Appl Ecol Environ Res 16:1935–1949

A Rocha Ghana, & IUCN (2019) Landscape Management Strategy Mole Ecological Landscape (2019–2029). A Rocha Ghana, KN3480, Kaneshie. Unpublished report, pp 86

A Rocha Ghana (2020) Mole Community-Led Woodland Restoration for Sustainable Wood Fuel Utilization. Baseline Biodiversity Survey Report. A Rocha Ghana, KN3480, Kaneshie. Unpublished report, pp. 44

A Rocha Ghana, & Noe (2022) Map of mole ecological landscape. KN3480, Kaneshie. Unpublished

Asare RA, Kyei A, Mason JJ (2013) The community resource management area mechanism: a strategy to manage African forest resources for REDD+. Phil Trans R Soc B 368:1–9

Breuer H, Lüdeke-Freund F (2014) Normative Innovation for Sustainable Business Models in Value Networks. In: Huizingh K, Conn S, Torkkeli M, Bitran I (eds) The Proceedings of XXV ISPIM Conference – Innovation for Sustainable Economy and Society, 8-11 June 2014, Dublin, Ireland. viewed 21 November 2022. Retrieved from https://sustainablebusinessmodel. org/2014/06/09/working-definitions-of-sustainable-business-model-business-model-for-sustainability/

Bull JW, Gordon A, Law EA, Suttle KB, Milner-Gulland EJ (2013) Importance of baseline specification in evaluating conservation interventions and achieving no net loss of biodiversity. Conserv Biol 28(3):799–809

Burkill HM (1985) The useful plants of west tropical Africa, viewed 20 February 2022. Retrieved from https://plants.jstor.org/stable/10.5555/al.ap.upwta.1_801

Butsic V, Kelly M, Moritz MA (2015) Land use and wildfire: a review of local interactions and teleconnections. Land 4(1):140–156

Catlin-Groves CL (2012) The citizen science landscape: from volunteers to citizen sensors and beyond. Int J Zoology 2012:1–14

CITES 2022 Caracal, viewed 18 February 2022. Retrieved from https://cites.org/eng/gallery/species/mammal/caracal.html

Dogbevi EK (2020) Special report: inside the illegal logging of rosewood in Ghana, viewed 11 February 2022. Retrieved from https://www.premiumtimesng.com/foreign/west-africa-foreign/376213-special-report-inside-the-illegal-logging-of-rosewood-in-ghana.html

Energy Commission (2006) Strategic National Energy Plan 2006-2020. Annex IV Report, viewed 25 June 2022. Retrieved from https://energycom.gov.gh/files/snep/WOOD%20FUEL%20final%20PD.pdf

Ernoul L, Wardell-Johnson A, Willm L, Bechet A, Boutron O, Mathevet R, Arnassant S, Sandoz A (2018) Participatory mapping: exploring landscape values associated with an iconic species. Appl Geogr 97:71–78

Forestry Commission (2004) A brief guide on the establishment of Community Resource Management Areas (CREMAs), viewed 11 October 2022. Retrieved from https://www.oldwebsite.fcghana.org/assets/file/Publications/Wildlife%20Issues/Complete%20Guide.pdf

Fugu M (2018) Ban on logging and charcoal production in Damango, viewed 11 October 2022. Retrieved from https://www.graphic.com.gh/news/general-news/ban-on-logging-and-charcoal-production-in-damango.html

Fugu M (2021) REGSEC closes down charcoal, Rosewood factories - In Savannah Region, viewed 11 October 2022. Retrieved from https://www.graphic.com.gh/news/general-news/ghana-news-regsec-closes-down-charcoal-rosewood-factories-in-savannah-region.html

Gann GD, McDonald T, Walder B, Aronson J, Nelson CR, Jonson J, Hallett JG, Eisenberg C, Guariguata MR, Liu J, Hua F, Echeverría C, Gonzales E, Shaw N, Decleer K, Dixon KW (2019) International principles and standards for the practice of ecological restoration. Restor Ecol 27: S1–S46. https://doi.org/10.1111/rec.13035

Hovis M, Cubbage F, Rashash D (2020) Designing a citizen science project for Forest landscapes: a case from Hofmann Forest in eastern North Carolina. Open J Forestry 10:187–203

Husseini R, Aboah DT, Issifu H (2020) Fire control systems in forest reserves: an assessment of three forest districts in the northern region. Ghana, Scientific African, p 7

ITTO (2021) Workshop explores key elements of successful forest landscape restoration in Asia-Pacific. International Tropical Timber Organization, viewed 9 October 2022. Retrieved from https://www.itto.int/news/2021/09/29/workshop_explores_key_elements_of_successful_forest_landscape_restoration_in_asia_pacific/

Kugbe JX, Fosu M, Tamene LD, Denich M, Vlek PLG (2012) Annual vegetation burns across the northern savanna region of Ghana: period of occurrence, area burns, nutrient losses and emissions. Nutr Cycl Agroecosyst 93(3):265–284

MLNR (2019) Enhancing local community engagement in natural resources management. CREMA Progress Report, Ministry of Lands and Natural Resources, viewed 21 November 2022. Retrieved from https://mlnr.gov.gh/wp-content/uploads/2019/07/Crema-Progress-Report.pdf

MOFA (2022) District directorate west Gonja. Ministry of Food and Agriculture, viewed 25 June 2022. Retrieved from https://mofa.gov.gh/site/sports/district-directorates/northern-region/250-west-gonja

Mole National Park. 2015. Mole National Park, viewed 11 February 2022. Retrieved from https://molenationalpark.org/index.php

Newton AC, Boscolo D, Ferreira PA, Lopes LE, Evans P (2018) Impacts of deforestation on plant-pollinator networks assessed using an agent based model. PLoS One 13(12):1–17

Obiri DB, Abukari H, Oduro KA, Quartey RK, Dawoe ELK, Twintoh JJ, Opuni-Frimpong E (2022) Rosewood (Pterocarpus erinaceus) as a de facto forest common for local communities in Ghana. Int J Biodivers Conserv 14(1):1–13

Osterwalder A (2013) A better way to think about your business model. Harvard Business Rev 6. viewed 20 February 2022. Retrieved from https://hbr.org/2013/05/a-better-way-to-think-about-yo

Poccard-Chapuis R, Plassin S, Osis R, Pinillos D, Pimentel GM, Thalês MC, Laurent F, de Oliveira Gomes M, Darnet LAF, Peçanha JDC, Piketty MG (2021) Mapping land suitability to guide landscape restoration in the Amazon. Land 10(368):1–23

Ruxton GD, Schaefer HM (2012) The conservation physiology of seed dispersal. Philos Trans R Soc Lond Ser B Biol Sci 367(1596):1708–1718

Schweizer D, van Kuijk M, Ghazoul J (2021) Perceptions from non-governmental actors on forest and landscape restoration, challenges and strategies for successful implementation across Asia, Africa and Latin America. J Environ Manag 286:112251

Stuart-Hill G, Diggle R, Munali B, Jo Tagg J, Ward D (2005) The event book system: a community-based natural resource monitoring system from Namibia. Biodivers Conserv 14:2611–2631

Stanturf JA, Kleine M, Mansourian S, Parrotta J, Madsen P, Kant P, Burns J, Bolte A (2019) Implementing forest landscape restoration under the Bonn challenge: a systematic approach. Ann For Sci 76(50):1–21

Stanturf JA (2021) Forest landscape restoration: building on the past for future success. Restor Ecol 29(4):1–5

Tigabu MS, Savadogo P, Oden PC, Xayvongsa L (2010) Enrichment planting in a logged-over tropical mixed deciduous forest of Laos. J For Res 21:273–280

Vihervaara P, Poikolainen L, Nedkov S, Viinikka A, Adamescu C, Arnell A, Balzan M, Lange S, Broekx S, Burkhard B, Cazacu C, Czúcz B, Geneletti D, Grêt-Regamey A, Harmáčková Z, Karvinen V, Kruse M, Liekens I, Ling M, Zulian G (2018) Biophysical mapping and assessment methods for ecosystem services. In: Enhancing ecosystem services mapping for policy and decision making. Technical Report, p 73

Williams BK, Brown ED (2014) Adaptive management: from more talk to real action. Environ Manag 53:465–479

Yaro JA, Tsikata D (2013) Savannah fires and local resistance to transnational land deals: the case of organic mango farming in Dipale, northern Ghana. Afr Geograph Rev 32(1):72–87

Wildlife Conservation Regulations (1971) viewed 18 February 2022. Retrieved from http://extwprlegs1.fao.org/docs/pdf/gha40817.pdf

The opinions expressed in this chapter are those of the author(s) and do not necessarily reflect the views of UNU-IAS, its Board of Directors, or the countries they represent.

Open Access This chapter is licenced under the terms of the Creative Commons Attribution-NonCommercial-ShareAlike 3.0 IGO licence (http://creativecommons.org/licenses/by-nc-sa/3.0/igo/), which permits any noncommercial use, sharing, adaptation, distribution and reproduction in any medium or format, as long as you give appropriate credit to UNU-IAS, provide a link to the Creative Commons licence and indicate if changes were made. If you remix, transform, or build upon this book or a part thereof, you must distribute your contributions under the same licence as the original. The use of the UNU-IAS name and logo, shall be subject to a separate written licence agreement between UNU-IAS and the user and is not authorised as part of this CC BY-NC-SA 3.0 IGO licence. Note that the link provided above includes additional terms and conditions of the licence.

The images or other third party material in this chapter are included in the chapter's Creative Commons licence, unless indicated otherwise in a credit line to the material. If material is not included in the chapter's Creative Commons licence and your intended use is not permitted by statutory regulation or exceeds the permitted use, you will need to obtain permission directly from the copyright holder.

Chapter 3
Community-Based Rangeland Restoration for Climate Resilience and Pastoral Livelihoods in Chyulu, Kenya

Josephat Mukele Nyongesa, Camila I. Donatti, and Tom Kemboi Kiptenai

3.1 Introduction

Socio-ecological production landscapes (SEPLs) are dynamic land use mosaics that have been shaped over the years by interactions between people and nature in ways that maintain biodiversity and provide people with goods and services needed for their well-being (Gu and Subramanian 2014, p. 1). Traditional ecological knowledge and governance structures have been useful natural resources management strategies to sustain SEPLs for provision of ecosystem goods and services across landscapes (Lee and Sung 2018, p. 92) and are recognised (Rist et al. 2010, p. 3) as a key baseline contributing to existing modern scientific resource management approaches. However, traditional approaches have been declining over time, compromising the provision of goods and services in SEPLs.

The Idaho Rangeland Commission report (2012), p. 3) defines rangelands as areas that are not farmed and mainly include grasslands, shrublands, woodlands, savannas, tundra, alpine, marshes and meadows, and deserts. Terrestrial rangeland ecosystems are generally characterised by low precipitation and are considered to be the world's largest ecosystem biome with high biodiversity and socio-economic and cultural value (Bengtsson et al. 2019, p. 1). Rangelands cover over 54% of the world's terrestrial surface (Rangelands Atlas 2021, p. 8) and support over 30% of world's human population (Sala et al. 2017, p. 467). Other than food and medicinal plants, rangelands provide vital provisioning services including water, pasture for

J. M. Nyongesa (✉) · T. K. Kiptenai
Conservation International (CI) Kenya, The Watermark Business Park Ndege Road Spring Court, Nairobi, Kenya
e-mail: jnyongesa@conservation.org

C. I. Donatti
Conservation International (CI), Arlington, VA, USA

© The Author(s) 2023
M. Nishi, S. M. Subramanian (eds.), *Ecosystem Restoration through Managing Socio-Ecological Production Landscapes and Seascapes (SEPLS)*, Satoyama Initiative Thematic Review, https://doi.org/10.1007/978-981-99-1292-6_3

livestock, and wildlife (Selemani 2020, p. 3864) and support livelihoods (Godde et al. 2020, p. 2). Carbon sequestration, crop pollination, and climate regulation are also main rangeland regulating ecosystem services (Ahlström et al. 2015, p. 895), while cultural services include recreation, aesthetic value, and traditional lifestyles. Soil formation and nutrient cycling, habitat, and biodiversity are essential supporting ecosystem services in rangelands recognised as the basis for the production of other ecosystem services (Baer and Birgé 2018, p. 3). Furthermore, Reicosky (Ed. Reicosky 2018) concludes that soil is the basis and source of most terrestrial ecosystem services. However, these types of ecosystems have been negatively impacted by degradation, compromising their ecological ability to effectively provide goods and services. Therefore, restoration and innovative management strategies can allow the continuous sustainable provision of rangeland ecosystem goods and services to sustain biodiversity and local community livelihoods (Eds. Gann and Lamb 2006, p. 1).

Kenya's rangelands are classified as Arid and Semi-Arid Lands (ASALs), covering about 89% of the country's land (Birch 2018, p. 2). Of Kenya's 47 administrative counties, 23 are in ASALs (ACAPS 2022, p. 1), and 36% of Kenya's population, 70% of its livestock herds, and 90% of its wildlife live in rangelands (Njoka et al. 2016, p. 11). The Chyulu landscape rangelands present an important socio-ecological production landscape (SEPL) because they provide distinct ecosystem services that sustain socio-economic development and maintain biodiversity for strengthened mutual human–nature interactions. The Chyulu landscape is a semi-arid zone that provides dry-season grazing reserves for wildlife, and pastoral and agro-pastoral communities' livestock. The livelihoods of over 2.5 million people in the largely pastoral communities of the Chyulu landscape are dependent on the rangelands (Opiyo et al. 2015, p. 298). The Chyulu landscape has gazetted protected areas: Chyulu, Amboseli and Tsavo National Parks, the Chyulu Forest Reserve, and local community group ranches, all of which are habitats for diverse wildlife species with potential for foreign exchange earnings. The Chyulu ecosystem is a habitat for African endemic "Big Five" animals—the elephant, African buffalo, lion, leopard, and black rhinoceros. Chyulu's area is approximately 420,000 ha and serves as a vital water tower for over three counties in the region and beyond, including Kajiado, Taita Taveta, Kilifi, and the coastal city of Mombasa (Fig. 3.1, Table 3.1). The Chyulu watershed has been recognised as a "fountain of life" (Mwaura et al. 2016, p. 46). Chyulu grasslands are crucial for local communities, especially for the livelihoods of indigenous pastoralists, fodder provision for livestock, biodiversity conservation for diverse flora and fauna species, and soil carbon sequestration.

Though Chyulu Hills Landscape as a SEPL has significant potential to provide a variety of vital ecosystem goods and services, its socio-ecological and economic sustainability is increasingly threatened by land degradation influenced both by anthropogenic and natural factors. Likewise, Kariuki et al. (2018, p. 47) concluded that sub-Saharan Africa's rangelands are experiencing pressures related to climate change and habitat destruction. Competing land uses, overgrazing, poor resource management, increasing demand for and overexploitation of natural resources,

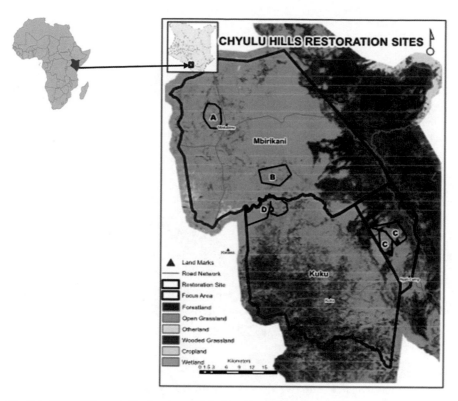

Fig. 3.1 Map of Kenya indicating geographical location of the Chyulu landscape, the study area (Source: GIS and Remote Sensing Vital Signs, CI)

logging, and charcoal production are evident in the Chyulu Hills Landscape. Declining Maasai traditional natural resource governance structures and limited management skills exacerbate the land degradation trend. Human factors are compounded by climate change events, mainly prolonged droughts, floods, unpredictable rainfall and seasonality patterns, and encroachment of invasive woody species on grasslands. The expansion of agricultural land due to an increasing human population has resulted in decreasing vegetation cover and encroachment into ecosystems such as wetlands (Pellikka et al. 2018, p. 178). Degradation continues to reduce the Chyulu rangelands' vegetation cover, in particular the grasslands' ecological capacity to provide ecosystem services, support biodiversity, and sustain human socioeconomic development. Vegetation cover reduction affects livestock and wildlife pasture and increases surface run-off and soil erosion, resulting in a reduction in soil organic carbon.

The Ecosystem-based Adaptation (EbA) approach and multi-stakeholder engagement that recognises multicultural indigenous knowledge and the value of nature can inspire rangeland natural resources management (NRM) and stimulate restoration actions for sustainable conservation of ecosystems in the face of climate change.

Table 3.1 Summary of case study sites in Chyulu Hills Landscape

Country	Kenya
Province/State	Rift Valley
County/District/Sub-County	Kajiado/Kajiado South
Size of Kajiado County (km^2)	21,900.9
Size of the Chyulu landscape (hectare)	410,533
Dominant ethnicity	Maasai
Target sites	Kuku "A", kuku "B" and Mbirikani group ranches
Size of three group ranches, the case study/project area (hectare)	151,205
Targeted restoration land size (hectare)	11,000
Number of direct beneficiaries (people)	1200
Number of indirect beneficiaries (people)	5000
Geographic coordinates (latitude, longitude)	Chyulu area: 2°58′37.9″S 37°46′30.2″E
Google map link for the study site	https://www.google.com/maps/d/edit?mid=1gM9L0 QeYcBlFNblUDS2QpLJDRfTSCmhb&usp=sharing; Copy of SITR-8-Kenya 1 – Google My Maps
Main economic activity	Pastoralism

Restored and managed SEPLs can sustain the provision of ecosystem services, conserve biodiversity, improve livestock productivity, and enhance the resilience of pastoralist communities to climate change. This approach can significantly reverse the land degradation trend. However, McGranahan and Kirkman (2013, p. 177) argue that sustainable rangeland management requires ecologically feasible strategies.

This chapter presents an assessment of the underlying ecological and socio-economic dynamics of Chyulu Hills Landscape to support the designing of restoration interventions to reverse rangeland degradation and manage ecosystems for provision of goods and services for sustainable development. The specific study objectives were as follows:

1. To assess land vegetation cover changes and carbon emissions.
2. To evaluate landscape degradation and land productivity.
3. To conduct a socio-economic feasibility assessment and design a rangeland restoration project.

The assessment was done as part of the Rangeland Restoration in Chyulu (RRC) project that aims to demonstrate that improved livestock and landscape management stimulates SEPL restoration, enhances carbon capture, builds climate-resilient ecosystems, and supports community livelihoods. The project applied the EbA strategy for climate change adaptation and an integrated natural resource management approach to generate sustainable socio-economic and ecological values across the Maasai community. The Chyulu Hills Landscape restoration case study is critically significant for SEPL management to sustain local communities' livelihoods and biodiversity conservation, which mainly depend on rangeland-grassland ecosystems.

3.2   Materials and Methods

3.2.1   Description of Study Area

The Chyulu Hills Landscape in South-eastern Kenya covers the three counties of Kajiado, Makueni, and Taita Taveta (Figs. 3.1 and 3.2). The study was conducted in three Maasai community group ranches (Mbirikani, Kuku "A", and Kuku "B") within Kajiado County. Kajiado County covers an area of 21,900.9 km^2, and is situated between Longitudes 360 5' and 370 5' East and Latitudes 10 0' and 30 0' South (GoK 2018, p. 16). The Chyulu Hills Landscapes are a volcanic mountain range about 150 km east of the Kenya Rift Valley (Scoon 2015, p. 1) and represent a vital ecosystem in the semi-arid Chyulu landscape. The landscape covers Chyulu Hills National Park, Tsavo West National Park, five community group ranches, and the Kibwezi forest reserve. The lower parts of the hills are composed of grasslands and woodland thickets, while area 1800 metres above sea level (m.a.s.l) is dominated by montane forest. The landscape is predominantly a rangeland covered with

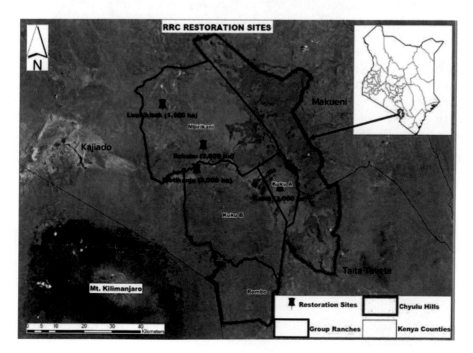

Fig. 3.2 Map showing the location of study sites (Source: GIS and Remote Sensing Vital Signs, CI)

Table 3.2 Group ranch summary

Group Ranch	Size in Hectares (Ha)	Average Registered Members	Population Estimates	Area (km²)	Targeted Community Beneficiaries	Annual Rainfall (mm)
Kuku "A"	18,712	1,996	37,141	1,280.3	5,000	350-500
Kuku "B"	9,600	5,516				
Mbirikani	122,893	4,600	18,617	1,923.4		
Total	151,205	12,112	55,758	3,203.7	5,000	

(Source: Ntiati 2002 and RRC project 2020)

grassland, important for livestock and wildlife pasture. The area has a bimodal rainfall pattern: October to December and between March and May. Annual precipitation ranges from 300 to 1250 mm. Temperature varies with altitude, ranging between 10 °C and 34 °C. The Chyulu Hills Landscape is characterised by climate change-induced long periods of drought and unreliable water sources. Pastoralism is the main economic activity. Main stocks are cattle, sheep, and goats. Pastoralism has been acknowledged in a recent study (Nyariki and Amwata 2019, p.1) as one of the main sources of livelihood in ASAL areas and contributors to Kenya's economy. Other economic activities include agriculture, tourism, ranching, quarrying, and small- and medium-sized commercial enterprises. The Chyulu ecosystem is an important water catchment, source of community livelihoods, and key biodiversity area for flora and fauna.

The Chyulu area is dominated by indigenous Maasai pastoralists (about 90%), and Kamba and other immigrant communities (around 10%). The three group ranches studied are approximately 151,205 ha in size (Table 3.2) and located at different coordinates. About 8.5% of Mbirikani (37.59°E; 2.51°S) is open grassland (Maclennan et al. 2009, p. 2). Kuku "A" and Kuku "B" are semi-autonomous group ranches, both referred to as "Kuku". The Kuku area is a crucial wildlife corridor between Amboseli and Tsavo National Parks. Kuku is located at 2°55′0″ S, 37° 40'60" E, at elevation of 1342 m.a.s.l. Previous studies (Kioko et al. 2006, p. 62) have reported the yellow fever tree (*Acacia xanthophloea*), Umbrella thorn (*Acacia tortilis*), and Blackthorn (*Acacia mellifera*) as the dominant tree species in the Chyulu Hills Landscape area.

3.2.2   Data Collection

Three prioritised group ranches were selected through purposive sampling based on multi-stakeholder participatory consultation and engagement through a stakeholder workshop, community agreements, the degradation status of the land where livelihoods are maintained, and restoration potential. Both qualitative and quantitative data on changes in vegetation cover, degradation, land productivity, and socio-economic status were collected. Secondary data on landscape degradation were collected through a literature review, and primary data on the socio-economic context were through 18 focus group discussions (FGDs), field observation, 12 key informant interviews, and field transect walks. Geographic Information System (GIS) and remote sensing and field transect approaches were used to collect natural capital and landscape degradation spatial and temporal data. For data analysis related to objectives 1 and 2, the Trends Earth tool, based on 250 m resolution MODIS data, was applied to track time series changes in land vegetation cover and carbon emissions from deforestation during the 2000–2019 period. Land use and land cover maps were developed from Landsat satellite imagery. Normalised Difference Vegetation Index (NDVI) was utilised to assess land productivity. Descriptive statistics were applied to analyse data for objective 3 and presented in the form of graphs and tables.

3.3 Results of Ecological and Socio-Economic Dynamics Assessment

3.3.1 Geographic Information System and Remote Sensing Results

3.3.1.1 Land Vegetation Cover Changes

Figures 3.3 and 3.4 illustrate land vegetation cover changes in the sites between 2000 and 2019. Areas covered with trees increased by 13.42% (55,093 ha), while grassland declined by 1.1% (4515.86 ha). A decline in grassland is linked to increasing climate change-induced drought and overgrazing. Wetland and cropland areas were

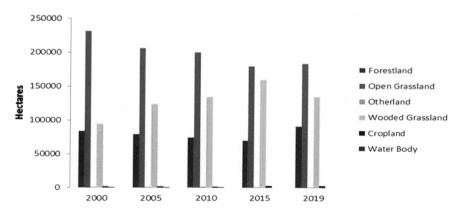

Fig. 3.3 Land cover change in Chyulu Hills Landscape between 2000 and 2019 (Source: GIS and Remote Sensing Vital Signs, CI)

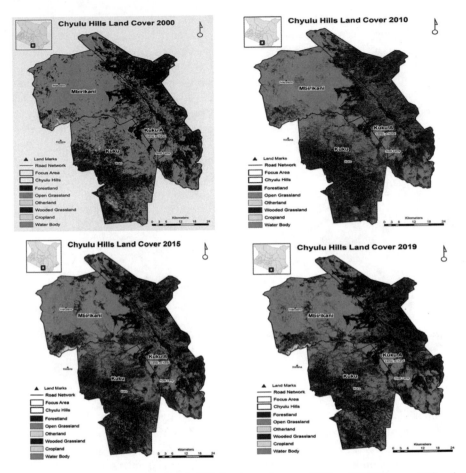

Fig. 3.4 Land cover changes in Chyulu Hills Landscape between 2000 and 2019 (Source: GIS and Remote Sensing Vital Signs, CI)

reduced by 0.66% (2709.5 ha) and 2.37% (9729.6 ha), respectively. Results corroborate the findings of Ehagi et al. (2018) that indicated reduction in land vegetation cover has extensive negative impacts on socio-economic development and ecological functions in Kajiado County.

Reduction in wetlands and croplands are closely correlated. Because agropastoralist farms are mainly rainfed, reduction in wetlands and long droughts influenced cropland expansion. For the purpose of this study, bushlands, open shrubland, and closed shrubland were clustered as tree-covered areas, and while they apparently show an increase, this does not imply that forested land increased. Decline in the grassland ecosystem has significant impacts on pastoral community livelihoods and biodiversity.

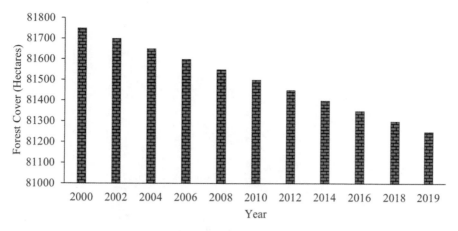

Fig. 3.5 Trend in forest loss in Chyulu Hills Landscape between 2000 and 2019 (Source: GIS and Remote Sensing Vital Signs, CI)

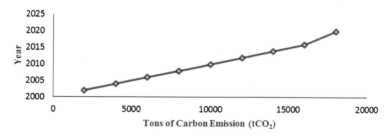

Fig. 3.6 Carbon emissions from deforestation in Chyulu Hills Landscape between 2000 and 2019 (Source: GIS and Remote Sensing Vital Signs, CI)

Trends in forest loss and carbon emissions between 2000 and 2019 are presented in Figs. 3.5 and 3.6, respectively. Results indicate declining (average 200 ha) forest cover in the three group ranches. This decline is linked to an increasing demand for forest products and agricultural land. A total of 70,472 metric tonnes of CO_2 was emitted relative to forest cover reduction. Increased forest loss peaked in 2019 relative to high carbon emissions.

The results show that decreased forest cover increased the volume of carbon emissions. Oduor et al. (2018, p. 1) observed that changes in land vegetation cover influence greenhouse gas emissions from landscapes due to deforestation practices. However, livestock enteric fermentation emissions across the Chyulu landscape were estimated at 32,000 tCO_2 /year. In the RRC project report, assessment of carbon sequestration potential from grassland restoration activities was projected to fall between 55,000 and 285,000 tCO_2 over a period of 5 years, based on the Verified Carbon Standards (VM0032) methodology.

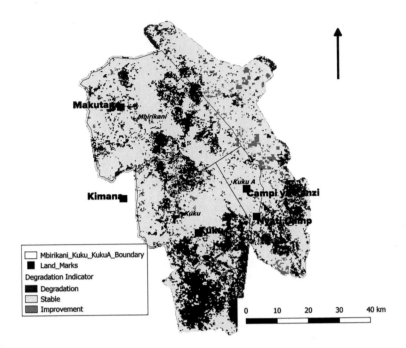

Fig. 3.7 Land degradation level in Chyulu Hills Landscape between 2000 and 2019 (Source: GIS and Remote Sensing Vital Signs, CI)

3.3.1.2 Landscape Degradation and Land Productivity (Fig. 3.7)

Approximately 27% (1,193.9 km^2) of land in the study sites was degraded, i.e. was either eroded, had reduced or no vegetation cover, or the land was overgrazed. Kuku ranches were highly degraded compared to Mbirikani. The variation in degradation is related to weak or lack of land management plans and the high livestock stock rate in Kuku compared to Mbirikani. The increased number of livestock herds in Kuku from outside the ranch compounded with a weak group ranch grazing management committee compared to Mbirikani also explains the variation in degradation levels. Land degradation impedes the function of the rangeland to provide ecosystem services for people's well-being and biodiversity conservation, as natural resources become scarce or no longer exist in degraded lands.

Normalised Difference Vegetation Index (NDVI) results revealed a declining land productivity trend (Fig. 3.8). Kuku group ranches had a higher level of decreased land productivity compared to Mbirikani. This variation is linked to the high degradation level in Kuku compared to Mbirikani. Degradation hinders the productivity of the grassland ecosystem and its capacity to provide pasture for wildlife, livestock, sequester carbon, and control soil erosion. A recent report on the effects of grassland degradation on soil quality and soil biotic community in a semi-arid temperate steppe (Han et al. 2020, p. 1) concluded that grassland degradation accelerates biodiversity loss and weakens ecological functions. The grassland

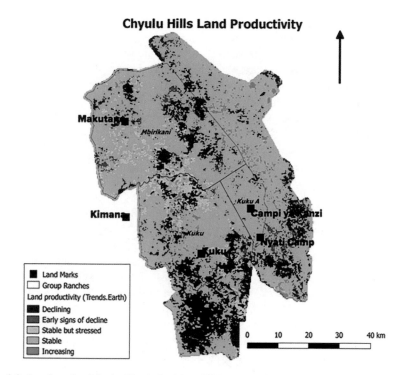

Fig. 3.8 Land productivity in Chyulu between 2000 and 2019 (Source: GIS and Remote Sensing Vital Signs, CI)

ecosystem is the main source of livelihood for indigenous pastoral communities in the Chyulu landscape, and therefore, land degradation exacerbates the reduced climate resilience of community livelihoods and ecological functions.

3.3.2   Socio-Economic Status of the Group Ranches

Socio-economic results were important for informed decision-making in designing restoration activities in the target areas. Communities in the three sites had similar socio-economic and cultural dimensions. The ranches are owned by the indigenous Maasai community with similar leadership structures. Two main organisations working within the sites are the Maasai Wilderness Conservation Trust (MWCT) and Biglife Foundation (BLF) in Kuku and Mbirikani, respectively. The high degradation recorded was linked to increasing loss of vegetation cover and pastures for livestock and wildlife in three sites. Decreasing rainfall and unpredicted seasons were common challenges reported in the group ranches. The sites are further characterised by an increasing human population, and hence, livestock overstocking lead to overgrazing. Colonisation by an invasive species, *Acacia oerfota*, known to suppress grassland, was recorded at the sites, though a recent report has established that the species has medicinal and nutritional values (Zarei et al. 2015, p. 2311). Livestock-wildlife competition for pasture was more common in Mbirikani than in Kuku. Most of the springs, swamps, and other wetlands were degraded, resulting in reduced water sources for humans, livestock, and wildlife.

Socio-economic factors that affected sustainable management of the rangelands included low awareness and skills on sustainable rangelands management practices, increasing human population, unplanned increase of the settlement areas, declining indigenous knowledge and traditional systems of governance, low technical capacity of group ranches, and inadequate gender mainstreaming in rangelands management.

3.3.2.1   Traditional Governance Structure for Natural Resources Management

The analysis showed that the Maasai community traditionally had a system of rotational pasture grazing and biodiversity conservation governance structures. Community leadership encouraged and practised *Oloopololi* (grass seed bank and grazing rotation) system, conserved pasture reserve for livestock during drought, and owned land communally. However, with changing socio-economic trends, indigenous traditional knowledge and governance mechanisms have declined over time, contributing to landscape degradation. The study also found that rangeland degradation was linked to livestock mismanagement, decreasing traditional ecological knowledge, and poor group ranch governance.

3.3.2.2   Actions Proposed by Communities to Reverse Landscape Degradation Threats

Through participatory FGDs, communities in the ranches proposed interventions to reverse ecosystem degradation, restore the landscape, and manage it sustainably (Table 3.3).

3.4   Designing of the Rangeland Restoration in Chyulu Hills Landscape

The assessment of the ecosystem, socio-economic, and ecological status of the landscape was used to guide the design for the RRC project for the targeted sites (Fig. 3.9).

The RRC project linked community development with restoration and improved rangeland natural resource management. Conservation International (CI) involved community group ranch leadership committees and two local NGOs in participatory

Table 3.3 Proposed landscape restoration interventions based on SEPL assessment and consultation with local communities

Landscape threat	Proposed interventions	Required/input
Ecological/ Conservation threats		
Soil/gully erosion	• Train community members to undertake restoration activities • Treat developing gullies to reduce soil erosion • Construct soil and water conservation structures • Establish community restoration incentives mechanism	Community incentives, Technical training, Restoration planting materials
Loss of vegetation cover	• Map degraded areas for restoration through re-seeding and assisted natural regeneration • Engage community scouts to protect restored areas • Establish women groups' grass seed banks	Incentives for casual workers, Planting materials
Invasive species	• Community sensitisation and training on pruning of such species to manage woody vegetation encroachment	Technical skills
Human-wildlife conflicts	• Engage community scouts to protect regenerating areas	Wages for casual workers
Socio-economic threats		
Low group ranch technical capacity and management skills	• Capacity building for ranch leaders on resource management and governance • Facilitate regular outreach community sensitisation engagements • Train grazing committees on sustainable grazing practices	Technical support, Community awareness campaigns
Increasing poverty	• Promote livelihoods enterprise diversification	Technical and financial support
Population increase and agricultural land expansion	• Develop community rangeland management plans	Technical skills
Declining traditional knowledge systems	• Engage ranch leadership to rebuild traditional resource management structures • Develop and disseminate resource management by-laws • Develop community conservation agreements	Grazing committee capacity building
Gender disparity	• Involve women and youth in restoration activities	Advocacy for gender inclusiveness

(Source: Authors' compilation, RRC progress reports 2022)

designing of the project activities. The consultative approach also brought together other key stakeholders in the landscape, including local administration, community opinion leaders, and government extension agents.

The joint participation enhanced landscape-wide involvement and commitment to sustainable restoration of degraded sites, while multi-stakeholder engagement built

Fig. 3.9 Inspection of degraded sites: Left: CI and BLF staff at Loosikitok; Right: CI and MWCT staff at Mortikanju (Photos: Josephat Nyongesa, 2021)

Fig. 3.10 Stakeholders meeting for Kuku "A" and "B" group ranch members (Photo: Agnes Nailantei, MWCT 2021)

the long-term resilience of the rangeland SEPL. A stakeholder workshop in January 2020 provided a platform to share study assessment results on the level of degradation, natural capital, carbon accounting pathways, and socio-economic contexts of the sites. The workshop was used to design the restoration strategy and endorse selected intervention sites. Workshop proceedings later guided the August–September 2021 inception meeting with CI partners, including community project validation and initiation of the project implementation framework (Fig. 3.10).

3.4.1 Overall Approach of the RRC Project

Though the target sites have been exposed to land degradation, the stakeholders' consultation, engagement, and landscape technical intervention strategic approaches have proven effective in reversing the degradation trend and restoring the degraded rangelands. The RRC project design was linked to an EbA approach. The Convention on Biological Diversity (2009, p. 9) defines EbA as an approach that involves the "use of biodiversity and ecosystem services as part of an overall strategy to help people adapt to the adverse effects of climate change". The RRC EbA interventions included vegetation cover restoration, soil and water conservation, and conservation and sustainable management of grassland ecosystems based on integration of indigenous knowledge and traditional governance structures with scientific expertise. Rangeland restoration based on EbA and natural resource management practices is a cost-effective approach with the potential to stimulate natural regeneration and large-scale socio-ecological and economic impacts in ASAL areas (Bourne et al. 2017, p. 7).

3.4.2 Multi-Stakeholder-Driven Rangeland Restoration Strategy

The participatory baseline study results guided the design of RRC project interventions. CI partnered with MWCT, BLF, and local community members to design and implement the RRC project activities. The community-driven strategic restoration design is structured along five objectives and expected outcomes, with key performance indicators to measure and assess progress (Table 3.4).

3.4.3 Progress and Achievements

3.4.3.1 Group Ranch Governance Capacity and Conservation Agreements

Ten community stakeholder engagements and capacity-building trainings on natural resource governance were conducted involving over 400 community members through community-based technical restoration managers and restoration officers who facilitated the day-to-day outreach activities. The technical team was trained on restoration techniques. Drafting of community conservation agreements with by-laws to conserve the four sites (Mbirikani: Loosikitok and Ilchalai; Kuku: Mortikanju and Kanzi) undergoing restoration has been initiated. The by-laws aim to contribute to rebuilding traditional resource management structures and practices such as *Oloopololi*. The agreements are voluntary and negotiated among community

Table 3.4 Rangeland restoration objectives, outcomes, key performance indicators, and activities

Objective	Outcomes	Key performance indicators	Activities
Goal: Demonstrate that improved livestock management and rangeland restoration can catalyse rangeland restoration, sequester carbon, and build climate-resilient pastoral livelihoods across the Chyulu landscape by 2025.			
1 Secure group ranch support for the intervention sites and build group ranch governance capacity.	• Institutional capacities and governance for ecosystem management improved • Conservation agreements developed	• Number of community conservation agreements endorsed • Number of community ranch members trained	• Engage community to develop by-laws regulating restoration site activities • Develop community conservation agreements
2 Strengthen the natural resource management skills of the group ranch.	• Empowered communities with skills in rangeland management • Restored rangelands that support improved landscape productivity and biodiversity	• Number of restoration plans developed • Number of women-led grass seedbanks established • Number of grazing scouts recruited and trained to protect restored sites • Number of restoration community crews recruited and trained	• Develop restoration plans • Recruit and train restoration casual workers • Train grazing scouts to manage livestock herds in restored areas
3 Undertake restoration interventions using livestock and restoration crews.	• Degraded areas restored • Rangeland best practices learned and shared	• Type of soil and water conservation technologies adopted • Degraded land restored (hectares) • Number of grazing plans developed • Number of people (casual workers) employed • Number of conservation incentive schemes established • Number of individual and households' beneficiaries	• Manage livestock in restored sites in accordance with restoration plans • Prune encroaching invasive species bushes • Source seeds from existing seedbanks • Reseed and rest degraded areas
4 Conduct carbon accounting on intervention sites and advance national climate objectives.	• Enhanced national capacity and knowledge sharing and learning on greenhouse gas emissions and reduction	• Number of dissemination products (peer-reviewed publications and policy briefs)	• Monitor carbon gains from restoration activities and account for estimated emissions reductions

(continued)

Table 3.4 (continued)

Objective	Outcomes	Key performance indicators	Activities
		• Amount in tonnes of carbon (tCO$_2$) sequestered	• Provide input to policy decisions, including matters on carbon rights and benefit-sharing frameworks
5 Build long-term rangeland restoration sustainability plan.	• Developed Chyulu rangeland integrated restoration management strategy	• Restoration strategy/plan	• Work with grazing scouts, group ranch leadership, NGOs, and local government partners to develop long-term sustainability restoration strategies

(Source: Conservation International, RRC project 2020)

members who commit to specified restoration and conservation interventions motivated by restoration incentives (casual restoration jobs, capacity building, and restoration scholarships). The aim is to enhance community well-being and NRM in exchange for keeping livestock outside restored areas and only grazing them according to a planned schedule. A total of 596 locals have been offered casual employment and are earning direct income that is currently benefitting about 1,800 household members. Two group ranch restoration committees were established to support management of the restoration casual workers. Two community education committees were also formed to oversee the RRC restoration scholarships incentive package, and 17 students from 17 schools were selected to receive scholarships. Every year, 34 high school students will be identified to benefit from a partial restoration scholarship (Fig. 3.11).

3.4.3.2 Strengthened Natural Resource Management Skills

Regular outreach community meetings have been organised to build community capacity on resource management and promote ecological literacy and restoration extension services. Over 500 community group ranch members were trained during the first 9 months of 2021. Development of four restoration plans that leverage livestock management to restore target sites has been initiated through community participation. Four women's groups (Inkisanjani, Lang'ata, Moilo, and Enkii) established grass seed bank restoration networks in Kuku to provide grass seeds for reseeding degraded sites and income generation for group members. Group ranches have identified 20 community scouts for training to support protection of restored areas. A total of 596 casual workers (Mbirikani: 312 for Loosikitok and 184 for Ilchalai; Kuku: 50 each for Mortikanju and Kanzi) were selected from households by group ranch committees and trained. The undertaking of restoration

Fig. 3.11 Students, parents, and teachers at induction meeting for the first cohorts of the RRC Scholarship from 17 schools in Kuku "A" and "B" ranches (Photo: MWCT 2021)

Fig. 3.12 Restoration training sessions for community-recruited casual workers in Kuku (L) and Mbirikani (R) (Photos: MWCT (L) and BLF (R), 2021)

Fig. 3.13 Restoration activities in Mortikanju: bund construction (L), and reseeding (M and R) (Photos: MWCT 2021)

activities has hence created local employment and generated income for community households. The project targets 5000 beneficiaries through NRM capacity building, restoration jobs, and strengthening of local enterprises. Figures 3.12 and 3.13 illustrate stakeholder restoration work and current intervention impacts.

3.4.3.3 Restoration of Degraded Rangeland Using Livestock and Restoration Crews

Communities' group ranch leaders were involved in the assessment process to identify appropriate site-specific restoration practices depending on level of site degradation. Trained casual workers implement restoration interventions, and expected socio-economic and ecological impacts are usually monitored before and after structural installations (Figs. 3.12, 3.13, 3.14, and 3.15).

Identified restoration interventions include construction of bunds or semi-circular micro-catchment Zai pits, stone lines, ponds, gully filling/healing, check dams, cut-off drains/ditches, sandbags, and contour grass strips. Constructed structures are complemented by grass reseeding and resting of degraded areas to stimulate assisted and natural vegetation regeneration correspondingly. A total of 8328 ha

Fig. 3.14 Restoration activities in Mbirikani: bunds construction in Ilchalai (L), constructed bunds in Loosikitok (R) (Photos: BLF 2021)

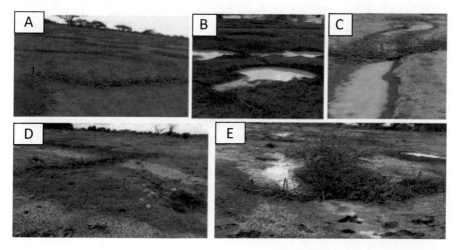

Fig. 3.15 Before and after restoration in Kuku-Mortikanju site: Above: bunds reinforced with local material barriers (**a**), stormwater, and silt trapped in bunds (**b**), and in gully (**c**). Below: grass regeneration in bunds and gully (**d** and **e**) (Photos: Josephat Nyongesa, 2021)

Table 3.5 Restoration work in Kuku and Mbirikani as of May 2022

Group ranch	Size (ha)	RRC sites	Targeted land for restoration (ha)	Land under restoration (ha)	No. of bunds constructed	No. of casual workers engaged
Mbirikani	122,893	Loosikitok	1,900	2,104	21,386	312
		Ilchalai/Olibili	2,600	1,724	15,191	184
Kuku "A"	18,712	Mortikanju	3,500	1,500	26,950	50
Kuku "B"	96,000	Kanzi	3,000	3000	0	50
Total	114,712		11,000	8,328	63,527	596

(Source: Authors' compilation, RRC progress reports 2022)

(as of June 2022) has been restored using different soil and water conservation structures, and rested and reseeded with grass species endemic to the sites, including Buffel grass (*Cenchrus ciliaris*), Masai love grass (*Eragrostis superba*), and Horsetail grass (*Chloris roxburghiana*). The project's target is to restore 11,000 ha by March 2024. A planned grazing scheme will be deployed to sustainably manage restored sites and avoid reversal of gains. Each group ranch restoration committee has initiated the process to recruit 100 women casual workers to gather and provide seeds for species endemic to the sites in addition to seed banks (Table 3.5).

3.4.3.4 Carbon Accounting

The RRC project adopted the Verified Carbon Standards (VCS) methodology for sustainable grassland restoration through a grazing regulation plan. The approach is being used to estimate atmospheric carbon removals from grassland restoration and greenhouse gas (GHG) emissions reduction from grazing management. Monitoring indicators aligned with national accounting systems for accounting GHG emissions and reductions have been developed. Indicators include trends in soil organic carbon content, soil bulk density, seasonal and long-term changes in herbaceous biomass, and grazing intensity. The indicators are being monitored throughout the project implementation phase. Local communities and implementing partners have been trained to collect the soil samples and biomass data. Lessons learned from carbon accounting shall provide input to policy decisions related to carbon and inclusion of soil carbon in Kenya's national policy decisions.

3.4.4   Future Challenges and Opportunities in the Restoration of the Chyulu Hills Landscape

The main challenges in the restoration of the Chyulu Hills Landscape rangeland ecosystem include limited and unpredictable precipitation, climate change, an increasing human population that exerts pressure on natural resources, lack of restoration plans and management capacity, group ranches land subdivision, limited funding for restoration, lack of or inadequate baseline information, and agriculture land expansion. However, increasing multi-stakeholder engagement in joint ecosystem restoration, community awareness on linkages between ecosystem degradation and their socio-economic well-being, and national and international commitments to mitigate climate change and restore ecosystems for livelihoods and biodiversity conservation present opportunities for future restoration efforts.

3.5   Discussion and Lessons Learnt

The Chyulu Hills Landscape provides abundant ecosystem services for people and conserves biodiversity. However, the ecosystem's socio-economic and ecological potential is under the threat of degradation driven by climate and human factors. This study presented key socio-economic and ecological assessment findings showing that the Kuku "A", Kuku "B", and Mbirikani group ranches in the Chyulu SEPL are experiencing degradation that is negatively affecting livelihoods and biodiversity conservation. The baseline assessment was important to assess ecosystem status to guide the design of restoration interventions. Its results were also valuable in enabling communities to understand linkages between natural value and their social well-being, as well as to identify landscape-wide multi-stakeholder restoration actions to reverse ecosystem degradation and enhance the management of the landscape to sustain socio-economic and ecological benefits. The local communities' traditional knowledge and natural resource governance structures, though declining because of changing social factors and economic diversification, still contribute immensely to the management of the landscape and were useful in designing the RRC project initiatives relative to the current social and conservation context.

The Maasai community's willingness to integrate their traditional natural resource management structures into new technological approaches was important for project sustainability and replication to other rangelands. The RRC project has demonstrated the potential and needs for landscape-wide multi-stakeholder consultation in joint participatory action to restore ecosystems. Ecosystem degradation can be reversed through multi-stakeholder engagement that enables a better understanding of human–nature connectivity and facilitates collective actions.

Building community leadership skills in resource management have shown that management of rangelands can contribute to ecosystem restoration and biodiversity conservation. Training community members on soil and water conservation yields quick restoration results, mainly vegetation regeneration to stop further land degradation and provide pasture for livestock. Planned grazing and livestock management is one of the key strategies essential for grassland restoration. Grassland in Chyulu contributes to soil carbon sequestration and provides pasture and habitat for rangeland biodiversity protection. Community capacity building strengthens restoration and management skills for sustainable ecosystem conservation. The RRC activities in target sites aimed to demonstrate that improved livestock management can catalyse rangeland restoration, sequester carbon, and build climate-resilient pastoral communities' livelihoods. The activities linked community development with improved natural resource management (NRM), restoration, and biodiversity protection.

The RRC project results have shown that restoration interventions can reverse the landscape degradation trend, improve grassland regeneration, rebuild indigenous ecological knowledge, improve group ranch natural resource governance, and enhance climate-resilient ecosystems and community livelihoods. Increased vegetation cover, control of soil erosion, community casual employment, and scholarships are some of the impactful incentives that motivate communities to restore and manage their SEPL. The timeframe of the RRC project is to end in March 2024. To date, interventions have established significant impacts on the restoration of this important SEPL, benefitting local people and improving ecological health for biodiversity benefits. Interesting observations from the RRC project include the increasing incidence of community members replicating learnt interventions on their own outside project sites, as well as multi-stakeholder collaboration on the commonly shared vision of restoring the landscape. Lessons from this case study can be replicated on a broader scale through rangeland restoration and improved land management by engaging landscape-wide multi-stakeholders.

Recognition of indigenous knowledge complemented the project's incentives to rejuvenate the landscape and motivated the communities' interest in restoring the degraded ecosystem. The local restoration casual jobs and support for youth education were critical social well-being incentives for the community to sustain conservation practices. Regular outreach and engagement contributed to rebuilding the communities' ecological literacy and livestock management skills. Training community paraprofessionals in soil and water conservation, livelihood opportunities, and participation in monitoring and evaluation of project gains inspired community members to act and participate in SEPL restoration and management.

Communities in the landscape depend on rangeland for their livelihoods. However, land degradation negatively impacts ecological functions affecting human socio-economic development and biodiversity. The RRC restoration interventions to date have demonstrated how degradation can be successfully reversed to restore a SEPL for human-nature interactive connectivity. Community conservation agreements and long-term restoration plans are essential strategic pillars bringing together different stakeholders in the landscape for sustainable SEPL restoration and management. Multi-stakeholder collective action is an essential restoration approach that contributes to reversing landscape degradation, creating socio-economic opportunities for people, and conserving biodiversity.

3.6 Conclusion

The communities' understanding of the indispensable value of the Chyulu landscape ecosystem to their livelihoods was the motivation behind multi-stakeholder efforts to restore and manage the degraded landscape for provision of ecosystem services, strengthening of climate resilience and local community well-being, and biodiversity conservation. This case study established the importance of a baseline assessment for informed decision-making in restoration project design. Decreasing land vegetation increased carbon emissions and accelerated ecosystem degradation. Landscape degradation reduces land productivity and affects sustainable provision of rangeland ecosystem services which support community well-being and biodiversity. By understanding the connectivity between nature's value and their socio-economic dependence on productive ecosystems, community group ranches brought together different stakeholders to actively participate in ecosystem restoration activities in the landscape. The RRC project focused on restoring rangeland productivity and considered indigenous traditional knowledge and governance structures to be useful in guiding restoration project design for SEPL management. Traditional knowledge integration in project activities is important for communities to relate their knowledge on restoration activities of degraded ecosystems. It is recommended to replicate similar case study interventions for scale.

Acknowledgements The authors thank Apple and CI for financial facilitation. We thank the editorial team for their guidance and paper review and peer reviewers for their useful comments which improved our paper. We recognise the implementing partners (MWCT: George Kingola, David Okul, Timthy Lenaiya, and BLF: Ernest Lenkoina and Jeremy Goss) for their cooperation and delivery on planned expected outcomes. We acknowledge CI Vital Signs for leading data collection, botanists Christopher Chesire and Mwadime Nyange and the Field Assistants from Kuku and Mbirikani. We thank the leadership of the group ranches for giving us the approval to carry out data collection. We are highly indebted to the Maasai communities for their cooperation and participation during the feasibility study and ongoing project implementation.

References

ACAPS (2022) Kenya: impact of drought on the arid and semi-arid regions. Assessment Capacities Project (ACAPS) thematic report, Kenya, viewed 26 April 2022, Retrieved from https://www.acaps.org/sites/acaps/files/products/files/20220331_acaps_thematic_report_kenyaimpact_of_drought.pdf

Ahlström A, Raupach MR, Schurgers G, Smith B, Arneth A, Jung M, Reichstein M, Canadell JG, Friedlingstein P, Jain AK (2015) The dominant role of semi-arid ecosystems in the trend and variability of the land CO2 sink. Science 348:895–899

Baer SG, Birgé HE (2018) Soil ecosystem services: an overview. In: Reicosky D (ed) Managing soil health for sustainable agriculture, vol 1. Burleigh Dodds Science Publishing, pp 1–24

Bengtsson JJ, Bullock M, Egoh B, Everson C, Everson T, O'Connor T (2019) Grasslands-more important for ecosystem services than you might think. Ecosphere 10(2):1–20

Birch I (2018) Economic growth in the arid and semi-arid lands of Kenya. Knowledge, Evidence and Learning for Development Report, 2018, Kenya, viewed 26 April 2022. Retrieved from https://assets.publishing.service.gov.uk/media/5c6fd72aed915d4a315f6552/482_Economic_Growth_in_the_Arid_And_Semi-Arid_Lands_of_Kenya.pdf

Bourne A, Muller H, Villiers AD, Alam M, Hole D (2017) Assessing the efficiency and effectiveness of rangeland restoration in Namaqualand, South Africa. Plant Ecol 218:7–22. https://doi.org/10.1007/s11258-016-0644-3

Ehagi D, Omuterema S, Nyandiko N (2018) Contributions of the agro-pastoral livelihoods to land cover change in Kajiado west Sub-County. Int J Sci Res Publ 8(6):254–258

Gann GD, Lamb D (Eds.) (2006) Ecological restoration: A mean of conserving biodiversity and sustaining livelihoods. Society for Ecological Restoration International and International Union for Conservation of Nature and Natural Resources, version 1.1, viewed 22 November 2021. Retrieved from https://cdn.ymaws.com/www.ser.org/resource/resmgr/custompages/publications/ser_publications/Global_Rationale_English.pdf

Godde CM, Boone RB, Ash AJ, Waha K, Sloat LL, Thornton PK, Herrero M (2020) Global rangeland production systems and livelihoods at threat under climate change and variability. Environ Res Lett 15(4):2

Government of Kenya (GoK) (2018) County Government of Kajiado, County Integrated Development Plan 2018-2022. Government Printer, Nairobi-Kenya, viewed 25 December, 2021. Retrieved from Kajiado County Integrated Development Plan 2018-2022.pdf (kippra.or.ke), https://repository.kippra.or.ke/bitstream/handle/123456789/995/Kajiado%20County%20Integrated%20Development%20Plan%202018-2022.pdf?sequence=1 & isAllowed=y

Gu H, Subramanian SM (2014) Drivers of change in socio-ecological production landscapes: implications for better management. Ecol Soc 19(1)

Han X, Li Y, Du X, Li Y, Wang Z, Jiang S, Li Q (2020) Effect of grassland degradation on soil quality and soil biotic community in a semi-arid temperate steppe. Ecol Process 9(63):1–11

Idaho Rangeland Commission (2012) Rangelands. An Introduction to Idaho's Wild Open Spaces. Rangeland Centre, University of Idaho, viewed 13 May 2022. Retrieved from Introduction-to-Rangelands-Launchbaugh.pdf (evergreen.edu)

Kariuki R, Willcock S, Marchant R (2018) Rangeland livelihood strategies under varying climate regimes: model insights from southern Kenya. Land 7(2):47

Kioko J, Okello M, Muruthi P (2006) Elephant numbers and distribution in the Tsavo-Amboseli ecosystem, south-western Kenya. Pachyderm 40:61–68. viewed 27 December 2021. Retrieved from https://www.awf.org/sites/default/files/media/Resources/Books%2520and%2520Papers/Pachy40.pdf

Lee HJ, Sung K (2018) Analysis of domestic and foreign local biodiversity strategies and action plan (LBSAP) using semantic network analysis. J Environ Impact Assess 27(1):92–104. https://doi.org/10.14249/eia.2018.27.1.92

Maclennan SD, Groom RJ, Macdonald DW, Frank LG (2009) Evaluation of a compensation scheme to bring about pastoralist tolerance of lions. Biol Conserv 142(11):2419

McGranahan DA, Kirkman KP (2013) Multifunctional rangeland in southern Africa: managing for production, conservation, and resilience with fire and grazing. Land 2(2):176–193

Mwaura F, Kiringe JW, Warinwa F, Wandera P (2016) Estimation of the economic value for the consumptive water use ecosystem service benefits of the Chyulu Hills landscape watershed, Kenya. Int J Agric Forestry Fisheries 4(4):36–48

Ntiati P (2002) Group ranches subdivision study in Loitokitok division of Kajiado District, Kenya. In: The land use change, impacts and dynamics project (LUCIDP)', working paper, no.7. Nairobi, Kenya. International Livestock Research Institute. viewed 26 December 2021. Retrieved from https://cgspace.cgiar.org/bitstream/handle/10568/1826/Lucid_wp7.pdf?sequence=1

Njoka JT, Yanda P, Maganga F, Liwenga E, Kateka A, Henku A, Mabhuye E, Malik N, Bavo C (2016) Kenya: Country situation assessment (working paper), viewed 23 December, 2021. Retrieved from https://idl-bnc-idrc.dspacedirect.org/bitstream/handle/10625/58566/IDL-58566.pdf?sequence=2&isAllowed=y

Nyariki DM, Amwata DA (2019) The value of pastoralism in Kenya: application of total economic value approach. Pastoralism Res Policy Pract 9(1):1–13

Oduor CO, Karanja N, Onwong'a R, Mureithi S, Pelster D, Nyberg G (2018) Pasture enclosures increase soil carbon dioxide flux rate in semiarid rangeland, Kenya. Carbon Balance Manag 13(24):1–13

Opiyo F, Wasonga O, Nyangito M, Schilling J, Munang R (2015) Drought adaptation and coping strategies among the Turkana pastoralists of northern Kenya. Int J Disaster Risk Sci 6:295–309

Pellikka PKE, Heikinheimo V, Hietanen J, Schäfer E, Siljander M, Heiskanen J (2018) Impact of land cover change on aboveground carbon stocks in Afromontane landscape in Kenya. Appl Geogr 94:178–189

Rangelands Atlas (2021) Rangelands Atlas Nairobi Kenya. ILRI, IUCN, FAO, WWF, UNEP and ILC, viewed 19 December 2021. Retrieved from https://www.rangelandsdata.org/atlas/sites/default/files/2021-05/Rangelands%20Atlas.pdf

Reicosky D (2018) Managing soil health for sustainable agriculture, Vol. 1 Fundamentals. Burleigh Dodds Science Publishing

Rist L, Uma Shaanker R, Milner-Gulland EJ, Ghazoul J (2010) The use of traditional ecological knowledge in forest management: an example from India. Ecol Soc 15(1):1–20

Sala OE, Yahdjian L, Havstad K, Aguiar MR (2017) Rangeland ecosystem services: Nature's supply and humans demand. In: Briske D (ed) *Rangeland Systems*. Springer Series on Environmental Management, pp 467–489. https://doi.org/10.1007/978-3-319-46709-2_14

Scoon R (2015) Geotraveller 21: Tsavo, Chyulu Hills and Amboseli, Kenya: Ancient landscapes and young volcanism (Geobulletin March 2015, Quarterly Publication of the GSSA), Academia. edu viewed 14 February 2022. Retrieved from https://www.academia.edu/en/62051043/Geotraveller_21_Tsavo_Chyulu_Hills_and_Amboseli_Kenya_Ancient_landscapes_and_young_volcanism_Geobulletin_March_2015_Quarterly_Publication_of_the_GSSA

Selemani IS (2020) Indigenous knowledge and rangelands' biodiversity conservation in Tanzania: success and failure. Biodivers Conserv 29(14):3863–3876. https://doi.org/10.1007/s10531-020-02060-z

Zarei M, Asgarpanah J, Ziarati P (2015) Chemical composition profile of wild acacia Oerfota (Forssk) Schweinf seed growing in the south of Iran. Orient J Chem 31(4):2311–2318

The opinions expressed in this chapter are those of the author(s) and do not necessarily reflect the views of UNU-IAS, its Board of Directors, or the countries they represent.

Open Access This chapter is licenced under the terms of the Creative Commons Attribution-NonCommercial-ShareAlike 3.0 IGO licence (http://creativecommons.org/licenses/by-nc-sa/3.0/igo/), which permits any noncommercial use, sharing, adaptation, distribution and reproduction in any medium or format, as long as you give appropriate credit to UNU-IAS, provide a link to the Creative Commons licence and indicate if changes were made. If you remix, transform, or build upon this book or a part thereof, you must distribute your contributions under the same licence as the original. The use of the UNU-IAS name and logo, shall be subject to a separate written licence agreement between UNU-IAS and the user and is not authorised as part of this CC BY-NC-SA 3.0 IGO licence. Note that the link provided above includes additional terms and conditions of the licence.

The images or other third party material in this chapter are included in the chapter's Creative Commons licence, unless indicated otherwise in a credit line to the material. If material is not included in the chapter's Creative Commons licence and your intended use is not permitted by statutory regulation or exceeds the permitted use, you will need to obtain permission directly from the copyright holder.

Chapter 4
Initiation of SEPLS Approach from World Peace Biodiversity Park (WPBP), Pokhara in Panchase Region of Nepal

Dambar Pun, Aashish Tiwari, Pabin Shrestha, Kapil Dhungana, Gunjan Gahatraj, and Dem Bahadur Purja Pun

4.1 Background on the Panchase Forest Conservation Area

Nepal is blessed with rich biodiversity. According to Ministry of Forest and Soil Conservation (MoFSC) (2014), Nepal is home to 6973 species of angiosperm and 26 species of gymnosperm. Koirala (1998) reported 306 plant species, including some medicinal plants from Panchase. Subedi (2002) reported 170 species of orchids around Pokhara Valley, of which 98 were from Panchase. The Panchase Protected Forest (PPF) hosts a total of 613 species under 393 genera and 111 families of flowering plants (Bhandari et al. 2018), and harbours unparalleled natural attractions and fantastic mountain landscapes (IUCN 2014; Neupane et al. 2021). In recognition of the rich biodiversity, forest resources, and cultural and spiritual values of the area, PPF was gazetted as a "Protected Forest", under Article 23 of the Forest Act in February 2012 (Suwal et al. 2013; DoF. 2012). Panchase, which literally means "five seats", is the meeting place of five hills, and is located in the Mid-Hills of Nepal (Baral et al. 2017). Subsequently, Protected Forests were renamed Forest Conservation Areas by the Forest Act of 2019.

Panchase region, a representative mid-mountain ecological zone of Nepal (Bhattarai et al. 2011), also functions as a river corridor between Chitwan National Park and Annapurna Conservation Area. Unlike other corridors, Panchase is a north-south corridor with great significance for wild animals during extreme climate

D. Pun (✉) · D. B. P. Pun
Back to Nature, Pokhara, Nepal

A. Tiwari · P. Shrestha
Conservation Development Foundation, Kathmandu, Nepal

K. Dhungana · G. Gahatraj
Ministry of Forest, Environment and Soil Conservation-Gandaki Province, Pokhara, Nepal

© The Author(s) 2023
M. Nishi, S. M. Subramanian (eds.), *Ecosystem Restoration through Managing Socio-Ecological Production Landscapes and Seascapes (SEPLS)*, Satoyama Initiative Thematic Review, https://doi.org/10.1007/978-981-99-1292-6_4

events. Community and national forests constitute the important forest ecosystem of Panchase. The forest ecosystem provides both provisioning and regulating services, such as soil erosion control, maintaining e-flow, and reducing siltation in Lake Phewa. This landscape supports rare, threatened, and endemic plant species, including six endemic and three threatened orchid species (Bajracharya et al. 2003; Subedi et al. 2007, 2011; Måren et al. 2014; Raskoti 2015; Bhandari et al. 2018). Four of these species—*Eria pokharensis, Gastrochilus nepalensis, Odontochilus nandae,* and *Panisea panchaseensis*—are local endemic (Bajracharya et al. 2003; Subedi et al. 2011; Raskoti 2015; Raskoti and Kurzweil 2015) and the other two—*Begonia flagellaris H. Hara* (Begioniaceae) and *Oberonia nepalensis* (Orchidaceae)—are region endemic and distributed in Central Nepal (Shakya and Chaudhary 1999; Rajbhandari and Adhikari 2009; Rajbhandari and Dhaugana 2010; Bhandari et al. 2018), including in and around the Panchase region. Beneficiaries of the different ecosystem services of the Panchase landscape range from those at the local level to the sub-national, national, and global levels (Bhandari et al. 2018).

4.1.1   Ecotourism in the Panchase Forest Conservation Area

Ecotourism as a component of the sustainable green economy is one of the fastest growing segments of the tourism industry due to its superiority compared to other types of tourism in terms of responsibility towards people, nature, and the environment (Poudel and Joshi 2020). Ecotourism has the potential to increase investment, jobs, exports, and technologies in the least developed and small island countries, whilst inflicting minimal adverse environmental impacts (Mazzarino et al. 2020). Since ecotourism is intrinsically linked with natural assets and cultural heritage, local and indigenous communities conserve and manage biodiversity and natural assets if they receive benefits. Practices that link local communities with the environment through ecotourism have played a significant role in mitigating climate change impacts and attaining development goals (K. C. and Thapa Parajuli 2014). Ecotourism supports environmental conservation and the sustainable development of affected communities (K. C. 2017).

The terrestrial ecosystem of the Panchase Forest Conservation Area (PFCA) consists of different land use types with forest covering 61% of the total area, followed by agriculture, grassland, and wetland covering 34%, 3%, and 1.3%, respectively (Adhikari et al. 2019). The upper region is managed primarily for the conservation of biodiversity and ecosystems. Pristine and dense forest provides provisioning services (i.e. fuelwood, wild fruits, medicinal and aromatic plants, and water), regulating services (i.e. water purification), and cultural services (i.e. recreation, aesthetics, and spiritual values) (Adhikari et al. 2019). The forest in the lowland is mainly managed as a community forest and is regarded as the fringe area of the PFCA (Adhikari et al. 2019). Panchase region provides habitats for 589 species of flowering plants, 24 species of mammals, and 262 species of birds, maintaining life cycles and genetic diversity (Bhandari et al. 2018). Maintaining the

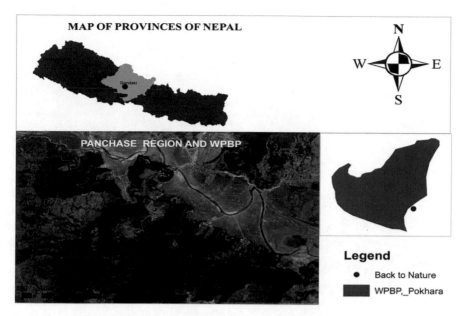

Fig. 4.1 Map of World Peace Biodiversity Park (WPBP), Pokhara along with Panchase Region in Gandaki Province, Nepal (Source: Prepared by authors using Arc GIS base map and polygon boundary based on Provincial Forest Directorate, Gandaki Province 2020)

landscape's integrity and heritage has allowed it to provide recreation and tourism opportunities for the roughly 3600 tourists and 25,340 pilgrims that visit every year (Bhandari et al. 2018). Mass tourism has brought changes to the lifestyle of the local people, lessening their adherence and attraction towards their own heritage. Indeed, the decline in consideration for conservation and protection of natural resources exhibited in their overuse in the PFCA has resulted in degradation of the environment, loss of economic benefits due to damage to resources or the local community, and disruption of local culture and values (Gautam 2020). Promoting ecotourism greatly emphasises raising the awareness of both communities and visitors, and allowing for equitable benefit sharing of tourism business with local communities.

4.2   Description of Study Area

World Peace Biodiversity Park (WPBP) is a small fractional portion of the Panchase Forest Conservation Area that is located 20 kilometres away from Pokhara, the capital of Gandaki Province (Fig. 4.1). WPBP contains a wide range of plant and animal species in *Schima-Castanopsis* forest, including *Schima wallichi*; *Castanopsis* species; differing orchid species; 35 identified bird species, like red junglefowl, kalij pheasant, and common dove; common leopard; black bear; barking deer; porcupines; snakes; and different species of insects like butterflies and fireflies.

Table 4.1 Brief description of the study area

Country	Nepal
Province	Gandaki
District	Kaski
Metropolitan city	Pokhara
Size of geographical area (hectare)	500
Dominant ethnicities	Gurung, Magar, Chettri, Dalit
Size of case study/project area (hectare)	18.85
Dominant ethnicity in the project area	Gurung, Magar, Chettri, Brahmin
Number of direct beneficiaries (people)	500 households
Number of indirect beneficiaries (people)	1,500 households
Geographic coordinates (latitude, longitude)	28°13'26.12"N, 83°54'10.72"E

The mosaic of land uses is characterised by agriculture, forestry, fisheries, animal husbandry, and tourism, which are the main economic activities of over 500 households of *Gurung, Magar, Chettri*, and *Dalit* ethnic communities from Nepal.

WPBP covers an area that is mostly part of the *Jauchare Dadhakarkha Bhirim* Community Forest, as well as the adjoining *Bhedikharka* and *Jhapulokhe Kharka* Community Forests. These community forests constitute the upstream area of Lake Phewa and are integral to the Lake Phewa basin due to the interacting livelihoods of both the upstream and downstream communities, and regulation of the e-flow of Lake Phewa. Although forests provide various goods and services for human well-being, the importance of ecosystem services arising from forests is not properly recognised in Nepal (Bhandari et al. 2018).

The WPBP area also falls under the Phewa Protected Watershed, designated by the Government of Nepal on 9 February 2022. Phewa watershed is a prime example of participatory watershed management in Nepal (Table 4.1).

4.2.1 Conservation Value and Challenges

Forests under PFCA are managed and operated by community forest user groups (CFUGs), which are legally entitled autonomous institutions, through operational plans that are technically supported by government and approved for implementation. These CFUGs carry out various activities including planting seedlings, controlling wildlife hunting, fighting forest fires, regulating grazing, monitoring forest encroachment, protecting the forests from natural hazards, and harvesting and regulation of the forest crop. However, the selection of high-value or preferred species, removal of old trees or unwanted species, heavy leaf litter collection, and illegal logging have detrimental impacts on the biological diversity and ecosystem function of community-managed forests (Shrestha et al. 2010).

The older generation has experienced and coevolved with the surrounding biodiversity, which has shaped attitudes, values, and beliefs related to biology, ecology, and use of the biodiversity (Popova 2014) in the WPBP area. However, younger generations are gearing towards off-farm income options, including tourism and foreign employment. This kind of generational economic shift has generated various issues, including the degradation of local and traditional knowledge on biological and cultural diversity, which has been a limiting factor behind the success of biodiversity conservation and ecotourism (Popova 2014). The PFCA is no exception in this regard. Outmigration amongst youth is substantial, resulting in labour shortages for forestry activities, farming, and livestock husbandry.

Issues such as dying water springs and water sources, invasive alien species, farmland abandonment, decrease in forest product availability, impacts of forest fires, and insect and disease outbreaks, are resulting in changes to the ecosystem services of Panchase region due to both natural and anthropogenic influences (Adhikari et al. 2019). Disturbance of wildlife due to mass tourism, haphazard village road construction, and encroachment on forests for constructing hotels and lodges, are other conservation challenges. Concerted efforts on the part of government authorities, local institutions, and local communities are of the utmost importance to address these issues and challenges to protect and enhance the conservation value of the PFCA, including WPBP.

4.3 Technical Approaches and Methods

The project adopted a participatory and iterative approach—employing a continuous cycle of action and reflection to ensure improved action. The participatory approach entailed bottom-up initiatives where local and indigenous communities were intensely involved in all activities. Collaborative work between indigenous people and external agencies is a key task towards improving the local and indigenous economy and relations with external actors, whilst also serving as a means to care for the environment across geopolitical boundaries (Popova 2014). The main vision behind WPBP, as stated in the management plan, lies in the prospect of promoting ecotourism based on biodiversity and cultural aspects whilst applying the principles of the Convention on Biological Diversity (CBD), which is also aligned with the socio-ecological production landscapes and seascapes (SEPLS) concept. SEPLS are the result of the interactions of nature and society in a given geographical and temporal context whose sustainability through time has depended on the careful management of ecosystem services and the multifunctional use of land within the carrying capacity and resilience of the environment. As a result, they constitute examples of areas with high capacity for biodiversity conservation, socioeconomic development, and preservation of cultural assets such as traditional knowledge and local traditions (Bélair et al. 2010). In light of the principles prominently embedded in the SEPLS concept, ecotourism has more benefits compared to adverse impacts on the environment, society, and culture.

4.3.1 Roles of Different Stakeholders

Back to Nature, a privately owned ecolodge established in 2011 with the aim of initiating conservation and sustainable use of biodiversity based on ecotourism in the Panchase landscape utilising local resources, values, customs, and folklore that foster the socio-economic well-being and resilience of the ecosystem. As a local entrepreneur, Back to Nature (Fig. 4.1) has made collective efforts to create an enabling environment for community participation, whilst providing support and lobbying on financial and technical aspects and coordinating during implementation of activities (Table 4.2). These were crucial for the transformation and reformation of the community forest into the politically recognised WPBP. Back to Nature provides and promotes ecotourism activities in the Panchase landscape, like trekking, jungle walks, bird watching, and homestays. It also supplements nature-based tourism activities like paragliding, rafting, and so on.

Table 4.2 Proposed activities of the WPBP Management Plan (Gandaki Province Forest Directorate, 2020)

No.	Activities	Planned Year				
		2020 (1st)	2021 (2nd)	2022 (3rd)	2023 (4th)	2024 (5th)
1.	Construction of welcome and entry gate	✓			✓	
2.	Construction and improvement of eco-trail	✓	✓	✓	✓	✓
3.	Plantation of *ficus* species (*bar* and *pipal*) for traditional resting place (*chautari*)	✓	✓			
4.	Information board, sign, and direction board, species ID and tags, website etc.	✓	✓	✓	✓	✓
5.	Biological control (invasive species management)	✓	✓	✓	✓	✓
6.	Ticket house		✓			
7.	Guardhouse		✓			
8.	Information centre		✓			
9.	Sanitation and toilet	✓	✓	✓	✓	
10.	Drinking water management	✓	✓			
11.	Orchid park	✓	✓	✓	✓	✓
12.	Enhancement of water sources		✓			
13.	Pond reconstruction		✓	✓	✓	

Community-based organisations like CFUGs, women's group, farmer's group, indigenous ethnic groups, street development committee, school teachers, and local communities, each have a niche in the ecotourism development and conservation priority activities of WPBP (e.g. school, youth club, and women's group were trained in education on biodiversity and ecotourism). Government organisations like the Provincial Ministry of Industry, Tourism, Forest and Environment; and line agencies like the Gandaki Province Forest Directorate, Panchase Protected Forest Office, the ward office, and others, have provided financial and technical support for formulation, implementation, and evaluation of activities, and provided government representation on the WPBP conservation committee.

4.3.2   Activities and Measures

The Provincial Ministry of Industry, Tourism, Forest and Environment, Gandaki Province provided technical support for the formulation and development of the WPBP Management Plan (2020–2024). This plan, primarily premised on biodiversity conservation and ecotourism development, charted out a set of activities for the 2020–2024 period. Activities are implemented by the conservation committee formulated under the WPBP user group and former community forest user groups, with representation from Back to Nature, the women's group, and other community-based organisations (Table 4.2). These activities are implemented by the WPBP conservation committee, which was transformed with an enhanced governance structure (Fig. 4.2).

4.4   Results

The results summarised below were obtained from implementation of the first year of activities in the WPBP management plan (Table 4.2). The WPBP conservation committee with coordination from Back to Nature aims to carry out ecotourism development and biodiversity conservation on the adjacent Panchase landscape through implementation of activities in the management plan based on a SEPLS approach mainstreaming with the principles of the CBD and ecotourism.

Visitors are mesmerised, educated, and healed when finding themselves near the lush green forest, waterfalls, and beautiful landscape, and exposed to the simple rural life and friendly people. Tourism activities, such as forest walks, trekking amidst lush green forests, dew walks, scenic mountain views, wild animal sightings, orchid treks, and firefly watching, greatly enhance the tourism experience of visitors. According to the local people, conservation of nature and biodiversity is important since they are dependent on nature for ecotourism. Households having an ecotourism business were found to have greater awareness about the significance of biodiversity conservation and the role played by ecosystem services, as well as awareness of

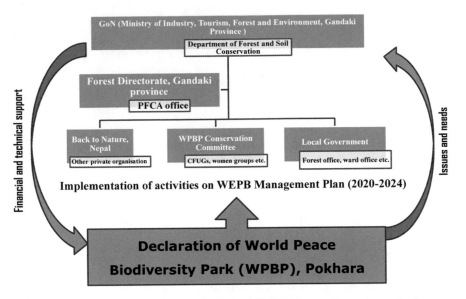

Fig. 4.2 Governance structure of the newly declared WPBP (Source: Prepared by author)

meetings and capacity building trainings held during formation of WPBP (Fig. 4.3). With the active leadership of Back to Nature alongside the generous support of the Government of Nepal and local communities, the desired results of conservation education and ecotourism development have been achieved to a great extent.

WPBP has been managed by a conservation committee advisory under the Ministry of Industry, Tourism, Forest and Environment, Gandaki Province. The conservation committee is mandated to take a leadership role in operating WPBP, as well as fundraising and financial management with potential national and international donors. The committee aims to publish information on the WPBP declaration and activities in national and local newspapers and online on social media platforms.

Impacts from society on the environment are to be minimised with the initiation of the SEPLS approach in the Panchase landscape and the development of ecotourism.

The ecotourism business operated in communities has put forth its best effort to provide authentic native and local cuisine, which offers unique tastes and promotes traditional knowledge and systems. This effort also prevents economic leakage, as most spending is retained in the local economy. As per the framework of the Global Sustainable Tourism Council (GSTC), sustainable ecotourism should benefit local and indigenous communities in an equitable manner. Equally important is educating visitors on local culture, heritage sites, and the significance of cultural assets in the promotion of ecotourism. Ecotourism has further motivated local and indigenous people to preserve and renovate cultural heritage sites such as monasteries, main walls, stone scriptures, and fresco paintings. WPBP was equipped with a welcome

Fig. 4.3 Laying foundation stone by the Provincial Forest Minister (Photo credit: Back to Nature, 2020)

gate leading to the entrance path. In addition, a trekking trail of approximately 300 m (2 m in width) was constructed to give visitors a quality trekking experience amidst the lush green forest and along a stream (Fig. 4.4). Due to a lack of effective coordination and support, the plan for a ticket house had to be moved to an upcoming year. The objective of these building activities was to add value to the ecotourism experiences of visitors and help generate benefits for the local and indigenous communities. Without benefits, local and indigenous communities will not take ownership of conservation and protection of these tourism assets.

Three rectangular *chautari* (traditional resting platforms) and one rectangular pond/wetland were constructed in the vicinity (Fig. 4.5). Customs, folklore, and traditional belief systems provided the primary factors for the design of these structures, which mirror traditional religious beliefs and local values. *Chautari* (traditional resting places) provide brief shelter for villagers, trekkers, and livestock herders on hot summer days. Oftentimes, local people gather in *chautari* to discuss the community agenda, but this trend has been weakening. Ponds are conserved and renovated to hold rainwater to the maximum extent. Natural ponds are used for irrigation, livestock drinking water, and recharging downstream villages and the sub-watershed. They have impervious clay or synthetic liners and engineered structures to control flow direction, liquid detention time, and water level (Kadlec and

Fig. 4.4 Eco-trail for initiation of ecotourism in WPBP (Photo credit: Back to Nature, 2020)

Zmarthie 2010). These structures can be vital to combat sedimentation, soil erosion, and the impacts of climate change on the Lake Phewa basin.

An information board was built and placed near the welcome gate (Fig. 4.6) to inform visitors of the objectives of WPBP and associated activities (Table 4.2). This information has improved the quality of the trekking experience for visitors. Three garbage buckets and an accessible toilet were also constructed along the eco-trail. These help to manage waste properly and keep the environment pollution free.

4.5 Discussion

The activities mentioned in the management plan include regulation of overharvesting and encroachment concurrent with the community forests of Nepal. The management plan was likewise oriented towards protecting water sources in the Phewa watershed for siltation control, with focus on services like drinking water and irrigation. Construction of *chautari* using *ficus* species, a traditional plant characterised by its long life, high branches, and adaptive features, was important from both biodiversity and sociocultural perspectives. In the context of participation, communities have shown a high level of active participation compared to the previous community forestry regime. This is due to motivation from government

Fig. 4.5 Constructed pond/wetland in WPBP Pokhara (Photo credit: Back to Nature, 2020)

commitment, indirect incentive generation, leadership development, and sustainable resource management from conservation and ecotourism-related activities in the management plan. The participatory approach with provisions for participation from women's groups, youth clubs, and schools, have contributed to traditional knowledge regeneration and conservation. The increasing trend of visitors to WPBP has also motivated youth, communities, and the private sector to engage in entrepreneurship (ecolodge), like Back to Nature.

Incentives provided by visitors and donor agencies will be utilised to further explore, initiate, and develop nature-based tourism activities like orchid and bird tourism, jungle walks, and butterfly watching in WPBP and the adjoining Panchase landscape. The vision behind the declaration of WPBP as a tiny portion of the whole Panchase landscape was to initiate a conservation and restoration programme through an endangered orchid park and protection programme for insects (especially fireflies and butterflies). Implementation of all activities in the management plan is to extend through 2024. The environment of participation, coordination, and commitment will provide the essential physical, social, and financial resources to initiate the orchid park programme based on protection, collection, and farming of orchid species in the park area. These synergies and trade-offs for in situ conservation and sustainable use through ecotourism can subtly be referred to as a "Living Laboratory on Ecotourism".

Fig. 4.6 Welcome gate of WPBP, Pokhara (Photo credit: Back to Nature, 2020)

Back to Nature, along with coordination from the conservation committee and community forests, is planning to initiate an eco-trail connecting the religious Shiva temple and Biodiversity Park to further enhance trekking experiences. Initiation of such public–private partnerships is innovative in itself, but challenges lie ahead, particularly on three fronts: (1) institutional capacity building of the conservation committee; (2) scaling up best practices out of all activities; and (3) ensuring financial sustainability through effective networking with potential national and international donor communities. Moreover, some pertinent issues, e.g. outmigration, human–wildlife conflict, forest fires, expansion of invasive alien plant species, and village road-induced landslides, must be addressed via concerted efforts and partnerships with government, NGOs, and donor communities. Finally, carrying capacity must be considered and prompt decisions made to address the issue of ecosystem restoration. Political instability and government organisational transformation[1] in the country have been a serious issue affecting the sustainability of WPBP, which relies on the government for technical and financial support. Another issue is operationalising of the conservation committee based on a sustainable flow

---

[1] Due to internal political conflicts in Nepal, government institutions like the Ministry of Industry, Tourism, Forest and Environment, Gandaki Province, were dissolved and the provincial ministry with power to regulate the PFCA and WPBP, Pokhara was changed to the Ministry of Forest, Environment and Soil Conservation, Gandaki Province.

of financial resources, which has been a delimiting factor for its sustainability and coordination effort in decision-making.

4.6 Conclusions

The declaration of WPBP in the Panchase landscape and implementation of various activities intrinsically linked to biodiversity and ecosystems have had positive impacts on restoring ecosystems, primarily with regards to: (1) forest and (2) water. Perhaps the best explanation of the correlation between ecotourism and ecosystem restoration can be found in the fact ecotourism spurred the local and indigenous people to confirm their stake in ecosystem restoration and made them aware that the very future of both ecotourism and the people themselves is reliant on a resilient ecosystem. WPBP in the Panchase landscape has created opportunities beyond the traditional resource management of community forestry with indirect use of cultural services and integration of local and scientific knowledge. It has provided incentives for conservation and restoration programmes in the degraded forest and agricultural ecosystem based on a SEPLS approach. However, these ecosystems are increasingly impacted by both climatic factors and anthropogenic threats. The financial and technical support of government agencies in the declaration process of WPBP and associated activities demonstrates that existing policy and legislative frameworks support the WPBP initiatives. The local and indigenous people of the Panchase landscape have a distinct and age-old relationship with the nature and biodiversity of the area that is essential for their survival and livelihoods. The activities of the management plan of WPBP have yielded positive results in terms of sustainable use of biological resources by adopting long-term sustainability, enhanced governance, and effective conservation of the SEPLS. Finally, further enhancement and perpetual succession of the WPBP and SEPLS concepts will require continued and effective coordination by Back to Nature along with the key stakeholders, including government authorities, local institutions, and donor communities.

References

Adhikari S, Baral H, Sudhir Chitale V, Nitschke CR (2019) Perceived changes in ecosystem services in the Panchase Mountain ecological region, Nepal. *Resources* 8(1):4

Bajracharya DM, Subedi A, Shrestha KK (2003) Eria pokharensis sp. nov. (Orchidaceae); a new species from Nepal Himalaya. J Orchid Soc Ind 17:1–4

Baral S, Adhikari A, Khanal R, Malla Y, Kunwar R, Basnyat B, Gauli K, Acharya RP (2017) Invasion of alien plant species and their impact on different ecosystems of Panchase area, Nepal. *Banko Janakari* 27(1):31–42

Bélair C, Ichikawa K, Wong BYL, Mulongoy KJ (2010). Sustainable use of biological diversity in socio-ecological production landscapes. Background to the 'Satoyama initiative for the benefit of biodiversity and human well-being

Bhandari AR, Khadka UR, Kanel KR (2018) Ecosystem services in the mid-hill forest of western Nepal: a case of Panchase protected forest. J Ins Sci Technol 23(1):10–17

Bhattarai KR, Måren IE, Chaudhary RP (2011) Medicinal plant knowledge of the Panchase region in the middle hills of the Nepalese Himalayas. Banko Janakari 21(2):31–39

DoF. (2012) Panchase protected forest management plan. Department of Forest (DoF). Ministry of Forest and Soil Conservation (MoFSC), Government of Nepal, Kathmandu

Gautam V (2020) Examining environmental friendly behaviors of tourists towards sustainable development. J Environ Manag 276:111292

IUCN (2014) Strengthening homestay business for diversifying livelihoods: building local people's resilience against climate change in the Panchase area. International Union for Conservation of Nature, p 4

K. C. A (2017) Ecotourism in Nepal. The Gaze J Tourism Hospitality 8:1–19

K. C. A, Thapa Parajuli RB (2014) Tourism and its impact on livelihood in Manaslu conservation area, Nepal. Environ Develop Sustain 16(5):1053–1063

Kadlec RH, Zmarthie LA (2010) Wetland treatment of leachate from a closed landfill. Ecol Eng 36(7):946–957

Koirala RA (1998) Botanical diversity within the project area of Machhapuchhre development organization, Bhadaure/Tamage VDC, Kaski district. A baseline survey for Machhapuchhre development organization (MDO). Bhadaure/Tamage, Kaski District, Nepal

Måren IE, Bhattarai KR, Chaudhary RP (2014) Forest ecosystem services and biodiversity in contrasting Himalayan forest management systems. Environ Conserv 41(1):73–83

Mazzarino JM, Turatti L, Petter ST (2020) Environmental governance: media approach on the United Nations programme for the environment. Environ Develop 33:100502

MoFSC. (2014) Nepal biodiversity strategy and action plan 2014–2020. Government of Nepal, Ministry of Forests and Soil Conservation, Kathmandu, Nepal

Neupane R, KC A, Aryal M, Rijal K (2021) Status of ecotourism in Nepal: a case of Bhadaure-Tamagi village of Panchase area. Environ Dev Sustain 23(11):15897–15920

Popova U (2014) Conservation, traditional knowledge, and indigenous peoples. Am Behav Sci 58(1):197–214

Poudel B, Joshi R (2020) Ecotourism in Annapurna conservation area: potential, opportunities and challenges. Grassroots J Nat Resour 3(4):49–73

Rajbhandari KR, Dhaugana SK (2010) Endemic flowering plants of Nepal, part 2. DPR Bull. Sp. Publication, Thapathali, Kathmandu

Rajbhandari KR, Adhikari MK (2009) Endemic flowering plants of Nepal, part 1. DPR Bull. Sp. Publication, Thapathali, Kathmandu

Raskoti BB (2015) A new species of Gastrochilus and new records for the orchids of Nepal. Phytotaxa 233(2):179–184

Raskoti BB, Kurzweil H (2015) Odontochilus nandae (Orchidaceae; Cranichideae; Goodyerinae), a new species from Nepal. Phytotaxa 233(3):293–297

Shakya LR, Chaudhary RP (1999) Taxonomy of Oberonia rufilabris (Orchidaceae) and allied new species from the Himalaya. Harv Pap Bot:357–363

Shrestha UB, Shrestha BB, Shrestha S (2010) Biodiversity conservation in community forests of Nepal: rhetoric and reality. Int J Biodiv Conserv 2(5):98–104

Subedi A, Chaudhary RP, Vermeulen JJ, Gravendeel B (2011) Panisea panchaseensis sp. Nov. (Orchidaceae) from Central Nepal. Nord J of Bot 29:361–365

Subedi A (2002) Orchids around Pokhara valley of Nepal. In: LI-BIRD occasional paper no.: 1. Local initiatives for biodiversity. Research and Development (LI-BIRD), Pokhara

Subedi A, Subedi N, Chaudhary RP (2007) Panchase Forest: an extraordinary place for wild orchids in Nepal. Pleione 1:23–31

Suwal RN, Bhuju UR, Tiwari KR, Pokhrel RK (2013) Preliminary identification of essential and desirable ecosystem services in the Panchase area of Nepal. A report environmental camps for conservation awareness (ECCA)/United Nations environment program (UNEP)

The opinions expressed in this chapter are those of the author(s) and do not necessarily reflect the views of UNU-IAS, its Board of Directors, or the countries they represent.

Open Access This chapter is licenced under the terms of the Creative Commons Attribution-NonCommercial-ShareAlike 3.0 IGO licence (http://creativecommons.org/licenses/by-nc-sa/3.0/igo/), which permits any noncommercial use, sharing, adaptation, distribution and reproduction in any medium or format, as long as you give appropriate credit to UNU-IAS, provide a link to the Creative Commons licence and indicate if changes were made. If you remix, transform, or build upon this book or a part thereof, you must distribute your contributions under the same licence as the original. The use of the UNU-IAS name and logo, shall be subject to a separate written licence agreement between UNU-IAS and the user and is not authorised as part of this CC BY-NC-SA 3.0 IGO licence. Note that the link provided above includes additional terms and conditions of the licence.

The images or other third party material in this chapter are included in the chapter's Creative Commons licence, unless indicated otherwise in a credit line to the material. If material is not included in the chapter's Creative Commons licence and your intended use is not permitted by statutory regulation or exceeds the permitted use, you will need to obtain permission directly from the copyright holder.

Chapter 5
Community-Based Restoration of Agroforestry Parklands in Kapelebyong District, North Eastern Uganda

Kizito Echiru and Samuel Ojelel

5.1 Introduction

In Uganda, 69% of households derive their livelihoods from subsistence farming (UBOS 2017). This reliance on agriculture coupled with a high population growth rate (3% per annum) has culminated in the clearance of natural landscapes. As a result, drylands (84,000 km^2, or 43% of the country), which have been identified as a Socio-ecological production landscape (SEPL) (Olupot 2015), have also been affected. According to Olupot (2015), this SEPL has undergone a large-scale conversion of land into agricultural farms and overexploitation of indigenous trees for wood fuel (charcoal and firewood). These human activities have resulted in the decimation of the traditional agroforestry parkland farming system (scattered trees in cultivated fields), a reduction in crop productivity, an increase in poverty, and the general degradation of environment (Olupot 2015). Common tree species include *Mangifera indica, Vitellaria paradoxa, Tamarindus indica,* and *Combretum* spp., among others. These are usually intercropped with millet, groundnuts, cowpeas, cassava, potatoes, rice, beans, maize, and green grams (Ojelel and Kakudidi 2015).

Trees play an integral part in the agroforestry parkland system by providing food, fuel, fodder, medicinal products, building materials, and saleable commodities, as well as contributing to the maintenance of soil fertility, water conservation, and environmental protection (Boffa 1999). Additionally, they contribute to improving the physical, biological, and chemical fertility characteristics of surrounding soils

K. Echiru
Save A Seed for the Future (SAFE), Soroti, Uganda

S. Ojelel (✉)
Save A Seed for the Future (SAFE), Soroti, Uganda

Department of Plant Sciences, Microbiology and Biotechnology, Makerere University, College of Natural Sciences, School of Biosciences, Kampala, Uganda

© The Author(s) 2023   77
M. Nishi, S. M. Subramanian (eds.), *Ecosystem Restoration through Managing Socio-Ecological Production Landscapes and Seascapes (SEPLS)*, Satoyama Initiative Thematic Review, https://doi.org/10.1007/978-981-99-1292-6_5

through a variety of processes. Data on the build-up of soil fertility and microclimate properties over the life of trees demonstrate that benefits are only realised over a long period. They also play a significant role in ensuring social equality, cultural stability, and spiritual values in traditional rural societies. More so, production of non-timber forest products in agroforestry parklands generates significant income for a variety of local economic actors. It, therefore, suffices to say that agroforestry parklands are pivotal in the attainment of the twin objectives of biodiversity conservation and sustainable livelihoods of communities. Owing to the importance of agroforestry parklands, Boffa (1999) opines that efforts to conserve and enrich existing parklands and to establish new ones are particularly critical.

This case study emanates from an action research project undertaken from 2019 to 2020 by Save a Seed for the Future (SAFE), Uganda (www.saveaseedforthefuture.org), with support from the Satoyama Development Mechanism (SDM). A baseline study was aimed at generating critical information regarding the reference ecosystem, a critical step in any restoration initiative. According to the Society for Ecological Restoration (SER), identifying a reference ecosystem is one of the eight principles of ecological restoration (Gann et al. 2019). The project targeted the sub-humid drylands of Kapelebyong District, Teso sub-region, where 91.9% of households are involved in subsistence crop farming (UBOS 2017). The project's objectives were to: (1) document tree diversity and traditional knowledge in remnant agroforestry parklands, (2) enhance community awareness and knowledge on agroforestry parklands, and (3) restore indigenous trees on 1500 hectares of cultivated fields. The project envisaged that the trees planted would provide timber and fuelwood; improve soil fertility and control water runoff; improve nutrition with fruits, nuts, and leaves; boost household incomes; and enhance habitat suitability and connectivity for different faunal species. Accordingly, the project would lead to the revitalisation of the traditional agroforestry parkland farming system, which would subsequently promote biodiversity conservation and sustainable livelihoods in harmony with nature.

5.2 Materials and Methods

5.2.1 Study Area

The project was conducted in the sub-humid drylands of Kapelebyong District, northeastern Uganda between 33°30'E to 33° 45'E and 2°24'N to 2°45'N (Table 5.1 and Fig. 5.1). Vegetation is predominantly savannah, dominated by *Combretum* species, *Vitalleria paradoxa*, and punctuated by seasonal as well as permanent streams. The area experiences a humid and hot climate with 1000–1350 mm of annual rainfall and temperatures ranging from 18 to 31.3 °C (Egeru 2012). Likewise, households involved in livestock farming represent 86.5%. The overall proportion of the population dependent on subsistence farming as the main source of livelihood is 86.2% (UBOS 2017). Crops grown in this area include

Table 5.1 Basic information of the study area

Country	Uganda
Province	Eastern
District	Kapelebyong
Municipality	-
Size of geographical area (hectare)	7,800
Dominant ethnicity(ies), if appropriate	Iteso
Size of case study/project area (hectare)	1,500
Dominant ethnicity in the project area	Iteso
Number of direct beneficiaries (people)	300
Number of indirect beneficiaries (people)	1,800
Geographic coordinates (latitude, longitude)	2°14'08.7"N 33°29'36.9"E- 2°25'00.7"N 33°43'57.1"E and 1°59'04.6"N 33°45'16.2"E- 1°58'49.7"N 33°53'35.5"E

millet, rice, sorghum, cowpeas, groundnuts, green grams, cassava, and sweet pota-toes (Ojelel and Kakudidi 2015). Livestock reared includes cattle, goats, and sheep, with farmers also relying on animal traction. The project activities were carried out in three of the seven sub-counties in Kapelebyong District, namely: Obalanga, Alito, and Okungur (https://kapelebyong.go.ug/lg/political-and-administrative-structure). The major tribes in this area are Iteso with small populations of Karamojong and Luo. The major ecosystems in the district include savannah woodlands, wetlands and river tributaries, and agricultural lands managed through private property, common property, and public property regimes. The district has two central forest reserves (public property), namely Akileng and Alungamosimosi, with the former located in Alito sub-county.

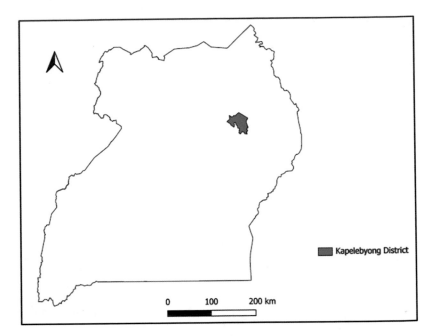

Fig. 5.1 Location of Kapelebyong District, northeastern Uganda (Base map drawn by Samuel Ojelel)

5.2.2 Implementation of Activities

Project activities commenced with a half-day stakeholder meeting on February 21, 2020, at Obalanga sub-county headquarters in the Kapelebyong District. Participants included among others District technical staff on natural resources, sub-county community development officers, parish chiefs, sub-county political leaders, representatives from the Iteso Cultural Union, and farmer group representatives (Fig. 5.2). The purpose of the meeting was to brief key stakeholders on project deliverables and to receive their feedback/insights. Thereafter, we commenced the botanical survey of tree species in the remnant agroforestry parklands with registered farmers as the target population. Twelve farmer groups were identified based on their interest in tree planting as expressed in their constitutions' aims and objectives. Each of the registered groups at respective sub-county community development offices has between 25 and 30 members.

Then, we randomly identified five subsistence farmers from each of the 12 registered farmer groups in the sub-counties of Obalanga, Okungur, and Alito in Kapelebyong District (totalling 60). We adopted the farmer group approach to facilitate mobilisation and target key informants with a preexisting interest in tree planting. A meeting was arranged with the group chairpersons to brief them on the research objectives, and five members were randomly selected from each group and thereafter interviewed in their respective villages (homesteads). The village is the

Fig. 5.2 Launch of project in 2019 at Obalanga sub-county headquarters, Kapelebyong District (Photo: Kizito Echiru)

lowest administrative unit in Uganda under the local government structure. We administered a semi-structured questionnaire to each farmer in their field (parkland). The questionnaire elicited information on the number and identity of trees per hectare, source of knowledge on tree preservation or planting on farm, benefits derived from trees, management of trees, and modes of harvesting tree products. The interviews took place in the parklands to minimise chances of missing tree species currently present or mistakenly including those absent.

Additionally, we conducted community sensitisation meetings (328 female and 209 male participants), produced and distributed brochures, and held a radio talk show on the Teso Broadcasting Service (87.6 FM). The brochures contained vital information on the value of trees on farms, common species in the agroforestry parklands of Kapelebyong, and management of trees on farms, among other information. This educational component was aimed at raising community awareness on the importance of trees on farms and tree planting in general.

We assisted two farmer groups with nursery bed equipment (wheelbarrows, spades, watering cans, trowels, potting paper) and assorted tree seeds namely *Grevillea robusta, Citrus sinensis, Psidium guajava, Eucalyptus camadulensis,* and *Markhamia lutea* (Fig. 5.3). These groups also received a stipend to promote motivation for raising the tree seedlings for the project. The groups were continuously monitored and supported by our field staff to ensure successful management of

Fig. 5.3 Community members pose with nursery bed equipment supplied by Save a Seed for the Future (Photo: Samuel Ojelel)

the nurseries. When the seedlings were ready for transplanting, the farmers prepared their fields and received the seedlings after field staff had ascertained the state of their fields.

5.2.3 Data Analysis

We computed the diversity of tree species in the agroforestry parklands using the Shannon-Wiener index based on the formula:

$$H' = \sum_{i-1}^{s} pi \ln pi$$

where p is the proportion (n/N) of individuals of a particular species (n) divided by the total number of individuals (N), and s is the number of species. Shannon's equitability (E_H) was calculated following Krebs[17] procedure, namely: dividing H by H_{max} (here $H_{max} = \ln S$) using the formula:

$$E_H = H/H_{max} = H/\ln S$$

Equitability assumes a value between 0 and 1 whereby 1 represents complete evenness. Means and percentages were used to analyse data on tree management strategies, tree harvesting techniques, and sources of knowledge on farm tree management. Graphs and tables were then used to present these results. The results of tree planting were analysed based on the survival rate of the seedlings distributed to the farmers and planted.

5.3 Results and Discussion

5.3.1 Documentation of Tree Species and Traditional Knowledge

5.3.1.1 Diversity of Tree Species

The study established an inventory of 43 tree species in 18 families in the agroforestry parklands of Kapelebyong District (Appendix 5.1). It was observed that only nine tree species (20.9%) were deliberately planted in the parklands, while the majority (79.1%) were retained when fields were first created. The species planted include *Mangifera indica, Anacardium occidentale, Carica papaya, Azadirachta indica, Melia azedarach, Eucalyptus camaldulensis, Psidium guajava, Grevillea robusta,* and *Citrus sinensis.* The planting of trees on cultivated fields increases species richness, diversity, and ecosystem services. A study by Buyinza et al. (2015) conducted among local communities in the Lake Kyoga basin recorded that the communities preferred preserving to planting trees on cultivated fields. This trend, however, contrasts with that reported in the agricultural landscapes of the Kigezi sub-region in western Uganda (Boffa et al. 2005). In the present case study, the weak tree planting ethos can be attributed to the perceived abundance of indigenous trees resulting in a lack of urgency to engage in tree planting.

The Shannon-Wiener diversity index of tree species in the agroforestry parklands of Kapelebyong is 3.23 with an equitability (evenness) of 0.86. This diversity index ($H' = 3.23$) is greater than 2.0, which denotes high diversity (Magurran 2004). The high equitability index shows that tree species are evenly distributed in the parklands. The mean tree density in these parklands is 5.7 trees ha^{-1}. This density is lower than the 15–42 mature trees/ha previously reported in some districts of Uganda (Byakagaba et al. 2011). Prevalent species include *Combretum adenogonium, Vitalleria paradoxa, Combretum collinum,* and *Mangifera indica* (Fig. 5.4). The presence of trees in the parklands underscores their ability to enhance resilience to environmental calamities such as climate change (Reppin et al. 2020). Based on prevalent tree species (Boffa 1999), the parklands in Kapelebyong can be described as *Combretum* spp., and *Vitalleria paradoxa* parklands.

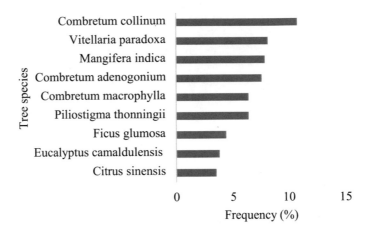

Fig. 5.4 The frequent trees in the agroforestry parklands of Kapelebyong District, Uganda (Echiru and Ojelel 2021)

Almost all tree species in these parklands are indigenous or naturalised. Only *Eucalyptus camaldulensis* and *Grevillea robusta* species were exotic species encountered. This trend emanates from the limited availability of exotic tree planting materials and the poor tree planting culture. The latter stems from the perceived relative abundance of indigenous trees and their associated products. However, elsewhere it has been reported that 69% of the species planted in Kigezi, western Uganda, were exotic (Boffa et al. 2005).

5.3.1.2 Uses of Parkland Trees

The farmers reported diverse uses of trees in the parklands (Fig. 5.5). The tree species provided 24 use categories with the greatest number of species contributing to provisioning services such as firewood, timber, and fruits (Fig. 5.5). This highlights the dependence of the community on the parklands as a livelihood buffer (Boffa 1999; Buyinza et al. 2015). In Burkina Faso, the carbon payment system promoted by the Reducing Emissions from Deforestation and Forest Degradation (REDD+) initiative is profitable for smallholder farmers who make efforts to plant and keep trees on farms (Neya et al. 2020). However, this programme is still in its infancy and has yet to benefit many smallholder farmers in Uganda. This signifies that parkland trees are exploited for their consumptive use value. This could be a precursor for non-renewability once demand exceeds supply.

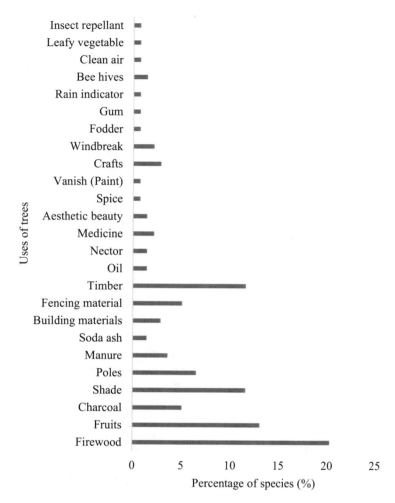

Fig. 5.5 Uses of agroforestry trees in Kapelebyong District, northeastern Uganda (Echiru and Ojelel 2021)

5.3.1.3 Knowledge of Preserving Trees in Parklands

Knowledge of preserving and/or planting trees on farms was acquired in many ways (Fig. 5.6). Most farmers learnt by imitating what their parents did when creating and managing crop fields. Knowledge was thus found to be passed from one generation to another through practical hands-on experience. However, given the increasing rate of urbanisation coupled with rural–urban migration, especially among the youths, it is feared that this transfer of knowledge will be increasingly impeded. The case study, therefore, makes a significant step towards ensuring the integration of traditional knowledge to provide prescriptions that are relevant to present scenarios. Figure 5.6 also shows that more farmers preserve trees on farmlands because of

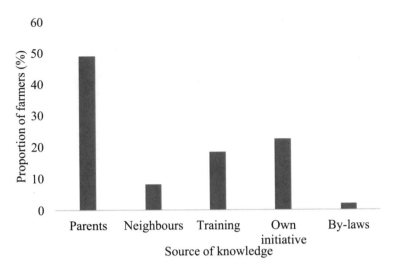

Fig. 5.6 Source of knowledge on tree preservation in the parklands of Kapelebyong District, Uganda (Source: Echiru and Ojelel 2021)

their own initiative and after trainings, aside from being obligated to do so according to by-laws. This provides a fundamental lesson to environmental practitioners to always prioritise awareness and sensitisation to create a change in the attitude of the target community. This data implies that people will protect species or ecosystems that are valuable to their livelihoods without necessarily being obligated to uphold laws. Therefore, these results affirm the assertion that "farmers are rational decision makers who choose to conserve and regenerate trees in their fields if this brings higher benefits" (Boffa 1999). The findings are akin to the policy mix approach whereby a combination of policy instruments that have evolved over time influence the quantity and quality of biodiversity conservation and ecosystem service provision (Ring and Schröter-Schlaack 2011).

5.3.1.4 Tree Management Techniques

The study documented three major tree management techniques, shown in Fig. 5.7. Weeding and pruning are the prevalent tree management practices in the parklands of Kapelebyong District (Fig. 5.7). Weeding is applied to control weeds when crops such as finger millet, ground nuts, and cowpeas are in the field, while pruning can be done with or without crops. Tree management through pruning offers twofold benefits: first, it reduces shade thus minimising tree competition with crops; and secondly, it provides harvestable materials such as firewood and construction materials, among others. Other tree management measures include spraying with insecticides (although not environmentally friendly), watering during the dry season, application of manure, coppicing, and pollarding. In West Africa, tree pruning is

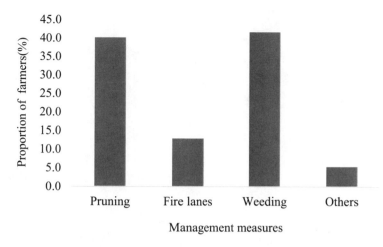

Fig. 5.7 Management of tree species in the parklands of Kapelebyong District, Uganda (Source: Echiru and Ojelel 2021)

an attractive option to improve crop production around tree canopies (Boffa 1999). Pollarding was not reported as a major tree management technique within the parklands of Kapelebyong. This can be attributed to limited use of parkland trees as fodder. According to Soule (1985), pollarding is carried out repeatedly by pruning branches at or near the same point, which results in the distinctive thick bushy appearance of the trees and is used for easing access for livestock, among other uses.

The tree management techniques aim at optimising the productivity of the parklands in terms of tree products and crop produce. The resultant benefits in turn act as an incentive for the farmers to engage in restoration of the parklands. As stated by Hemida and Adam (2019), effective management of trees on farms (parklands) acts as a key livelihood strategy to increase and diversify income sources and strengthens farmers' ability to improve their livelihoods. Thus, tree management techniques are applied according to the needs of the farmers.

5.3.1.5   Harvesting Techniques

We recorded six techniques applied to harvesting trees and tree products in the parklands of Kapelebyong District (Fig. 5.8). The most prevalent method is pruning (especially of branches and twigs) while debarking is the least applied. The low prevalence of debarking and digging roots points to the prudence of farmers and limited knowledge on the use of tree barks and roots. This demonstrates that farmers are judicious when harvesting trees on their farms. It also points to an in-built culture of ensuring sustainable use of the available resource. Buyinza et al. (2015) opine that the most important species will suffer the greatest harvesting pressure from local communities. The harvesting techniques documented in this case study could pose variable effects on the trees depending on the parts harvested. For example,

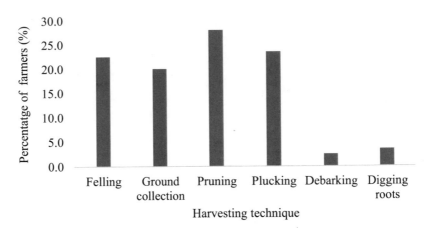

Fig. 5.8 Techniques of harvesting tree products in the agroforestry parklands of Kapelebyong District (Source: Echiru and Ojelel 2021)

harvesting of fruits has been reported to cause a significant impact on regeneration and population viability (Gaoue and Ticktin 2008 cited in Delvaux et al. 2009) while harvesting of the bark or roots lessens tree survival (Vermeulen 2006, Delvaux et al. 2009). Therefore, a prerequisite for sustainable management of agroforestry trees is knowledge on the response of species to harvesting techniques. However, the current study did not investigate the differential impacts of harvesting for each tree species recorded in the parklands.

5.3.2 Raising Community Awareness on Agroforestry Parklands

We held five community sensitisation meetings, distributed brochures, and broadcast a radio talk show on 89.7 FM Teso Broadcasting Service to raise community awareness on the value of agroforestry parklands. The meetings reached out to 537 (328 females and 209 males) participants in the sub-counties of Alito, Obalanga, and Okungur. Awareness motivates farmers and deepens their knowledge and understanding of the importance of agroforestry parklands. As observed by Tesfaye et al. (2014), it also influences farmers' decision-making on conservation actions.

5.3.3 Restoring Tree Cover in Cultivated Fields

We distributed at least 14,500 tree seedlings to 300 farmers in the 12 groups for planting on their individual farms (Figs. 5.9 and 5.10). At the time of project closure,

Fig. 5.9 Potted tree seedlings (Photo: Kizito Echiru)

Fig. 5.10 A farmer after
receiving tree seedlings
(Photo: Kizito Echiru)

the survival rate was 75% (10,875). Factors affecting this survival rate include destruction by termites and stray livestock, especially goats, and erratic rains. It was evident that farmers with large pieces of land opted to dedicate their fields to tree planting in the form of plantations and woodlots rather than agroforestry parklands. Trees planted were *Grevillea robusta, Citrus sinensis, Psidium guajava, Eucalyptus camadulensis*, and *Markhamia lutea*. These species were selected due to their adaptability to local environmental conditions, agroforestry value such as having open crowns for light penetration, multi-purpose production capacity (timber, firewood, fruits), fast growth rates, and acceptability among farmers. Although the project aimed to restore indigenous species, this could not be fully achieved due to scarcity of planting material and slow growth rates. Cognisant of these obstacles, we reached a compromise to utilise both native and exotic tree species, provided they met the criteria explained earlier.

The success of the restoration efforts can be monitored in several ways. First, there is a need for continuous field monitoring and support visits by the implementing agencies, local government extension workers, and funding agencies. This would help to bridge technical gaps as well as recommend any corrective actions that need to be undertaken. Secondly, the stakeholders can conduct joint evaluations of the progress and achievements of the restoration efforts using prior determined indicators such as tree survival rates. Thus, every restoration project ought to have an elaborate results-based strategy for monitoring and evaluation of its impacts.

5.4 Conclusion, Lessons, Challenges, Opportunities, and Recommendations

5.4.1 Conclusions

This study recorded 43 tree species in the parklands in Kapelebyong District with a high species diversity index and evenness. The trees provide diverse products to the farmers. Most of the trees are indigenous species preserved during opening of the land. The project raised community awareness, which was instrumental in enhancing community knowledge on agroforestry parklands and participation in raising and planting at least 14,500 tree seedlings in their farmlands towards restoring agroforestry parklands. The engagement of the community in prioritisation of the tree species and participation in raising seedlings allowed the community to take ownership of the restoration initiative.

5.4.2 Lessons, Challenges, and Opportunities

The lessons learnt during execution of this project include the following: (1) the community willingness to restore what is beneficial to their livelihoods (local relevance of restoration or any other interventions); (2) the importance of traditional knowledge in developing reference (baseline) information, for example, species lists, upon which restoration activities can be anchored; (3) the need to strike a win-win situation when choosing between native and exotic tree species to be prioritised in restoration initiatives; and (4) the need to take advantage of the environmental challenges and resource scarcity facing communities to create opportunities for garnering support for restoration.

We could not meet the demands and aspirations of the entire community in Kapelebyong District due to the small resources at our disposal. The UN (2019) opined that ecosystem restoration is costly in terms of time and financial resources with approximately 2000 USD required to restore 1 hectare (UN 2019). We also faced the challenge of finding and raising indigenous tree seedlings. In fact, there is a knowledge gap on this matter at the national level. This gap is partly responsible for the preference of exotic over native tree species in tree planting initiatives (Kalema and Hamilton 2020). Additionally, we faced challenges from unpredictable weather patterns (droughts and rains), insects (especially termites), and stray livestock (mostly goats and cattle), which destroy tree seedlings thereby reducing the survival rates. Another challenge encountered centred on indigenous versus exotic tree species. The farmers tend to prefer the latter due to their fast growth as opposed to the former. Additionally, the continued dependence on biomass energy (firewood and charcoal) is a serious impediment to restoration efforts.

The restoration of ecosystems can benefit from increasing environmental awareness and cautiousness in the community. This is occasioned by the visible scarcity of valuable resources such as firewood, timber, construction materials, and so forth. Environmental practitioners can galvanise support around scarcity of vital resources to rally communities to undertake ecosystem restoration. Globalisation has also accelerated information sharing, networking, and collaboration, which can be harnessed to design and implement ecosystem restoration projects. Additionally, there are national and international commitments and declarations for the UN Decade on Ecosystem Restoration 2021–2030 (UN 2019), which can help to mobilise support and financial resources for ecosystem restoration projects. There are also funding agencies offering funding on a competitive basis for projects geared towards ecosystem restoration, for example, the Global Environment Facility (GEF).

5.4.3 Recommendations

Basing on the experience from this case study, we recommend the following:

(a) The knowledge gap regarding the propagation of native tree species needs to be filled to avail indigenous tree seedlings in a timely manner.
(b) There is a need to sensitise the populace on the benefits of native tree species over exotic species, but a combination of the two may be a more realistic option.
(c) There is a need to put in place measures that ameliorate factors that decrease tree survival rates, for example, adoption of environmentally friendly anti-termite measures, soil moisture conservation measures such as mulching during dry spells and limiting the movement of stray livestock.
(d) There is a need to understand the impacts of different tree harvesting techniques on regeneration, population viability, and tree survival in agroforestry parklands.
(e) There is a need to scale up the restoration of agroforestry parklands in the drylands of Uganda to optimise the twin objectives of biodiversity conservation and sustainable livelihoods.

Acknowledgements We extend our gratitude to the Institute for Global Environmental Strategies (IGES) through the Satoyama Development Mechanism (SDM) for the financial support towards the implementation of this project. We are also grateful to the local government staff and the farmers in the Kapelebyong District who actively participated during the conceptualisation and implementation of the project.

Appendix 1 Agroforestry parkland tree species and their uses in Kapelebyong District, Uganda

Family	Scientific name	Local name (Ateso)	Utilisation
Anacardiaceae	*Anacardium occidentale* L.	Ekasnat	Fr, Oi, Fi
	Mangifera indica L.	Emiebe	Fr, Ma, Sh, Wb, Fi, Cr
	Ozoroa insignis Delile	Etiling	Ti, Po, Fi, Cr
	Sclerocarya birrea (A.Rich.) Hochst.	Ejikai	Ti, Fr
Annonaceae	*Annona senegalensis* Pers.	Ebwolo	Cr, Fr, Sh
Apocynaceae	*Carissa spinarum* L.	Emuriei	Fr, Fm, Me
Caricaceae	*Carica papaya* L.	Epapalu	Fr
Combretaceae	*Combretum adenogonium* Steud. ex A.Rich.	Emeng	Po, Fi, Sh, Ca
	Combretum collinum Fresen	Ekulony	Po, Fi, Sh, Bm
	Combretum macrocalyx (Tul.) Jongkind	Ekoboi	Fi, Sh, Ch, Wb, Bh, Ti, Po,
	Combretum molle R. Br. ex G. Don	Ekwooro	Fi, Ch
	Terminalia superba Engl. & Diels	Ekokobot	Po, Ch, Fi, Ma
Euphorbiaceae	*Bridelia scleroneura* Müll. Arg	Erieco	Fr, Po, Fi
Fabaceae	*Acacia hockii* De Wild	Ekisim	Fi, Fm

(continued)

Family	Scientific name	Local name (Ateso)	Utilisation
	Albizia coriara Oliv.	Eteka	Ti, Fi
	Albizia zygia (DC.) J.F.Macbr.	Ebata	Ti, Wb, Fi
	Erythrina abyssinica Lam.	Engosororoi	Fi, Ab, Fm, Ra
	Philenoptera laxiflora (Guill. & Perr.) Roberty	Ekaikai	Fo, Fi
	Piliostigma thonningii (Schumach.) Milne-Redh.	Epapai	Pa, Fi, Bh, Sh, Sa, Hu
	Tamarindus indica L.	Epeduru	Fr, Me, Fi, Ch, Ab, Sp
	Acacia sieberiana (DC.) Kyal. & Boatwr.	Etirir	Fi, Fm, Ti
	Senegalia senegal (L.) Britton	Ekodokodoi	Fm, Ti, Po
Lamiaceae	*Vitex doniana* Sweet	Ekwarukei	Fr, Fi, Ma, Ti
Lauraceae	*Persea americana* Mill.	Ovacado	Fr, Sh.
Malvaceae	*Grewia mollis* Juss.	Eparis	Fr, Po, Fi, Ch
Meliaceae	*Azadirachta indica* A.Juss.	Abach	Me, Fi, Ir
	Melia azedarach L.	Elira	Ti
	Pseudocedrela kotschyi (Schweinf.) Harms	Eputon	Ti, Bm
Moraceae	*Ficus thonningii*	Emidit	Fi, Sh, Ma, Fi, Fr
	Ficus glumosa Del.	Ebiong	Ma, Ti, Sh, Fr
	Ficus platyphylla Del.	Ebule	Gu, Ma, Sh, Ti, Fr
	Ficus sycomorus L.	Eboborei	Sh, Ti, Fi
	Artocarpus heterophyllus Lam.	Efene	Fr, Sh
Myrtaceae	*Eucalyptus camaldulensis* Dehnh.	Ekalitusi	Ti, Fi
	Psidium guajava L.	Emapara	Fr, Sh
Proteaceae	*Grevillea robusta* A.Cunn. ex R.Br.	Egrivellia	Ti, Fi
	Protea madiensis Oliv.	Ebalangait	Fi, Sh, Sa
Rubiaceae	*Gardenia ternifolia* Schumach. & Thonn.	Ekoroi	Fm, Po
	Mitragyna stipulosa (DC.) Kuntze	Eutdolei	Fi, Sh
Rutaceae	*Citrus sinensis*	Emucuga	Fr
	Harrisonia abyssinica Oliv.	Ekerei	Ne, Bm, Cr, Fm
Sapotaceae	*Vitellaria paradoxa* C.F.Gaertn.	Ekungur	Oi, Ch, Fr, Fi, Ne
Zygophyllaceae	*Balanites aegpytiaca* (L.) Delile	Ecomai	Lv, Fi, Ch

KEY: *Fi* Firewood, *Fr* Fruits, *Ch* Charcoal, *Sh* Shade, *Po* Poles, *Ma* Manure, *Sa* Soda ash, *Bm* Building materials, *Fm* Fencing materials, *Ti* Timber, *Oi* Oil, *Ne* Nectar, *Me* Medicine, *Ab* Aesthetic beauty, *Sp* Spice, *Pa* Paint for fish nets, *Cr* Crafts, *Wb* Windbreak, *Fo* Fodder, *Gu* Gum, *Ra* Rain indicator, *Ca* Clean air, *Lv* Leafy vegetable, *Ir* Insect repellent
Source: Echiru and Ojelel, 2019
Source: Echiru and Ojelel 2021

References

Boffa JM (1999) Agroforestry parklands in sub Saharan Africa. In: FAO conservation guide 34. Food and Agriculture Organization of the United Nations, Rome

Boffa JM, Turyomurugyendo L, Barnekow-Lillesø JP, Kindt R (2005) Enhancing farm tree diversity as a means of conserving landscape-based biodiversity. Mt Res Dev 25(3):212–217

Buyinza J, Agaba H, Ongodia G, Eryau K, Sekatuba J, Kalanzi F, Kwaga P, Mudondo S, Nansereko S (2015) On-farm conservation and use values of indigenous trees species in Uganda. Res J Agric Forestry Sci 3(3):19–25

Byakagaba P, Eilu G, Okullo JBL, Tumwebase SB, Mwavu EN (2011) Population structure and regeneration status of *Vitellaria paradoxa* (C.F. Gaertn.) under different land management regimes in Uganda. Agric J 6(1):14–22

Delvaux C, Sinsin B, Darchambeau F, Van Damme P (2009) Recovery from bark harvesting of 12 medicinal tree species in Benin, West Africa. J Appl Ecol 46(3):703–712

Echiru K, Ojelel S (2021) Utilization and population structure of tree species in the agroforestry parklands of Kapelebyong District, north eastern Uganda. Res J Agric Forestry Sci 9(3):1–8

Egeru A (2012) Role of indigenous knowledge in climate change adaptation: a case study of the Teso sub-region, eastern Uganda. Indian J Tradit Knowl 11(2):217–224

Gann GD, McDonald T, Walder B, Aronson J, Nelson CR, Jonson J et al (2019) International principles and standards for the practice of ecological restoration. Restor Ecol 27(S1):S1–S46

Gaoue OG, Ticktin T (2008) Impacts of bark and foliage harvest on *Khaya senegalensis* (Meliaceae) reproductive performance in Benin. J Appl Ecol 45(1):34–40

Hemida MAA, Adam YO (2019) The importance of farm trees in rural livelihoods in eastern Galabat locality, Sudan. Agric Forestry J 3(2):81–88

Kalema J, Hamilton A (2020) Field guide to the forest trees of Uganda: for identification and conservation. CABI

Magurran AE (2004) Measuring biological diversity. Blackwell, Oxford

Neya T, Abunyewa AA, Neya O, Zoungrana BJ, Dimobe K, Tiendrebeogo H, Magistro J (2020) Carbon sequestration potential and marketable carbon value of smallholder agroforestry parklands across climatic zones of Burkina Faso: current status and way forward for REDD+ implementation. Environ Manag 65:203–211

Ojelel S, Kakudidi EK (2015) Wild edible plant species utilized by a subsistence farming community in Obalanga sub-county, Amuria District, Uganda. J Ethnobiol Ethnomed 11(1):7

Olupot W (2015) SEPLS definition and issues – a case of Uganda's drylands. In: UNU-IAS & IGES (ed) *Enhancing knowledge for better management of socio-ecological production landscapes and seascapes (SEPLS)*, Satoyama initiative thematic review, vol 1, pp 79–89

Reppin S, Kuyah S, de Neergaard A, Oelofse M, Rosenstock TS (2020) Contribution of agroforestry to climate change mitigation and livelihoods in Western Kenya. Agrofor Syst 94(1): 203–220

Ring I, Schröter-Schlaack C (2011) Instrument mixes for biodiversity policies. Helmholtz Centre for Environmental Research

Soule J (1985) Glossary for horticultural crops. John Wiley and Sons

Tesfaye A, Negatu W, Brouwer R, Van der Zaag P (2014) Understanding soil conservation decision of farmers in the Gedeb watershed, Ethiopia. Land Degrad Develop 25(1):71–79

Uganda Bureau of Statistics (UBOS) (2017) The national population and housing census 2014: area specific profile series, Kampala, Uganda

United Nations (2019) The United Nations Decade on Ecosystem Restoration, viewed 30 May 2022. Retrieved from https://wedocs.unep.org/bitstream/handle/20.500.11822/31813/ ERDStrat.pdf?sequence=1&isAllowed=y

The opinions expressed in this chapter are those of the author(s) and do not necessarily reflect the views of UNU-IAS, its Board of Directors, or the countries they represent.

Open Access This chapter is licenced under the terms of the Creative Commons Attribution-NonCommercial-ShareAlike 3.0 IGO licence (http://creativecommons.org/licenses/by-nc-sa/3.0/igo/), which permits any noncommercial use, sharing, adaptation, distribution and reproduction in any medium or format, as long as you give appropriate credit to UNU-IAS, provide a link to the Creative Commons licence and indicate if changes were made. If you remix, transform, or build upon this book or a part thereof, you must distribute your contributions under the same licence as the original. The use of the UNU-IAS name and logo, shall be subject to a separate written licence agreement between UNU-IAS and the user and is not authorised as part of this CC BY-NC-SA 3.0 IGO licence. Note that the link provided above includes additional terms and conditions of the licence.

The images or other third party material in this chapter are included in the chapter's Creative Commons licence, unless indicated otherwise in a credit line to the material. If material is not included in the chapter's Creative Commons licence and your intended use is not permitted by statutory regulation or exceeds the permitted use, you will need to obtain permission directly from the copyright holder.

Chapter 6
Farmland Management Strategies by Smallholder Farmers in the Mount Bamboutos Landscape in Cameroon

Louis Nkembi, Tankou Christopher Mubeteneh, Asabaimbi Deh Nji, Ngulefack Ernest Forghab, and Njukeng Jetro Nkengafac

6.1 Introduction

The Bamboutos Mountains are a group of volcanoes based on a swell in the Cameroon Volcanic Line shared by three administrative regions of Cameroon, merging in the north with the Oku Volcanic Field (Burke 2001). The large volcanic complex extends in a NE-SW direction for over 50 km, with the highest peaks rising to more than 2600 metres above sea level (m.a.s.l.) around the rim of a caldera with a diameter of 10 km. The upper part of the massif above 2000 m.a.s.l has a wet climate with more than 2500 mm of annual rainfall. The Bamboutos mountain landscape represents a key watershed. It gives rise to several rivers and lakes across the country, including the Mbam tributaries and Mifi tributaries (Ewane et al. 2021). Mt. Bamboutos is a biodiversity hotspot with a high degree of bird endemism and biodiversity. Some very important species of biodiversity endemic to this ecosystem include the primate Preuss's guenon (*Cercopithecus preussi*), Cooper's mountain

L. Nkembi · A. D. Nji · N. E. Forghab
Environment and Rural Development Foundation (ERuDeF), Civil Society Building, Buea, Cameroon

T. C. Mubeteneh
Department of Crop Sciences, Faculty of Agronomy and Agricultural Sciences, University of Dschang, Dschang, Cameroon

N. J. Nkengafac (✉)
Environment and Rural Development Foundation (ERuDeF), Civil Society Building, Buea, Cameroon

Institute of Agricultural Research for Development (IRAD) Ekona Regional Research Centre, Buea, Cameroon
e-mail: officevpsrts@erudef.org

© The Author(s) 2023
M. Nishi, S. M. Subramanian (eds.), *Ecosystem Restoration through Managing Socio-Ecological Production Landscapes and Seascapes (SEPLS)*, Satoyama Initiative Thematic Review, https://doi.org/10.1007/978-981-99-1292-6_6

squirrel (*Paraxerus cooperi*), the banded wattle eye (*Platysteira laticincta*), and Bannerman's Turaco (*Tauraco bannermani*), as well as green monkeys. Other species include the endangered Cross River gorilla (*Gorilla gorilla diehli*) and the Nigeria-Cameroon chimpanzee (*Pan troglotes ellioti*). This ecosystem also plays host to viable populations of species from many taxa, especially insects, plants, reptiles, amphibians, and small mammals.

Besides its unique biodiversity, this mountain remains the only ecosystem in the country cutting across three administrative regions, namely West, South-West, and North-West, encompassing over 20 villages with a human population between 20,000 and 30,000 (Fig. 6.1, Table 6.1). Most of these people depend on the mountain and its biodiversity for their livelihood; they practice slash-and-burn and clear the forest for farmland, leading to high rates of deforestation, destruction of water catchments, and loss of biodiversity and soil fertility (Tankou et al. 2013). Due to population pressure, farming is carried out on the steep slopes, leading to erosion and loss of soil nutrients. Cattle grazing is a common practice on the upper slopes where food crop cultivation is uneconomical (Yerima and Van Ranst 2005). The destruction of almost all catchments has led to serious water shortages. Population pressure on limited land has resulted in encroachment into marginal sloping areas, causing continuous erosion and regular landslides (Fig. 6.2).

Intensification of agriculture and horticulture (Fig. 6.3) in the Mt. Bamboutos landscape has led to soil erosion, poor soil quality, and food and water contamination, and has resulted in decreased yields, reduced incomes, and food insecurity (Abubakar et al. 2020). Farmers use high levels of fertilisers and pesticides for horticultural crops and have a poor understanding of the real economic and environmental costs—hence the high risk of increased soil and agroecosystem degradation and declining crop yields. Poverty remains widespread and is strongly associated with rural livelihoods (Andrianarison et al. 2022). Toh et al. (2018) carried out a study on the socio-economic impact of land use and land cover change on the inhabitants of the Mt. Bamboutos Caldera of the Western Highlands of Cameroon and concluded that a majority of households were poor and lived in abject poverty.

Morphological, physical, and chemical studies on the soils of Mt. Bamboutos showed that the soils of this region can be divided into seven groups according to the US Soil Taxonomy: lithic dystrandept soils, typical dystrandept soils, oxic dystrandept soils, typical haplohumox soils, typical kandiudox soils, tropopsamment soils, and umbriaquox soils. These soils are organised into three main categories: soils with andic characteristics in the upper region of the mountain (lithic dystrandept soils, typical dystrandept soils, and oxic dystrandept soils); ferralitic soils in the lower part of the mountain (typical haplohumox soils and typical kandiudox soils), and imperfectly developed soils (tropopsamment soils and umbriaquox soils) (Tematio et al. 2004).

The soils are characterised by low bulk density (0.73 g/cm^3) and a loamy texture with the low bulk density indicative of the andosolic nature of these soils, which might be due to more ground biomass input in the form of leaves. The soils' highly fine particle (silt + clay) content is associated with the absence of translocation of

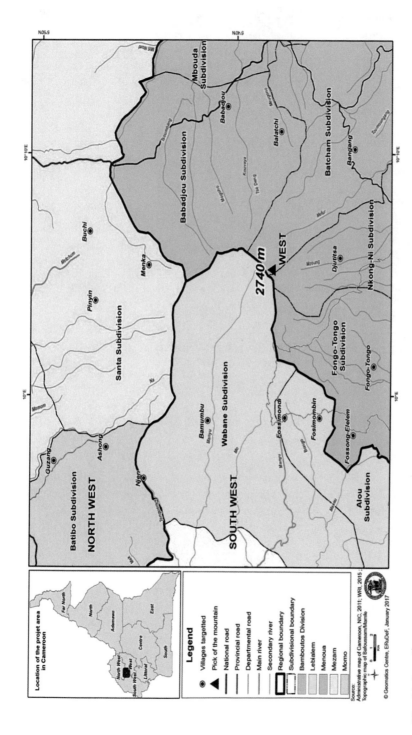

Fig. 6.1 Location map of the project area showing the targeted villages (Source: ERuDeF Geomatics Centre)

Table 6.1 Basic information of the study area (Source: Mount Bamboutos Initiative (MBI))

Country	Cameroon
Regions	South west, north west, and west regions
District	Bamboutos, Menoua, Mezam, Momo, and Lebialem
Municipality	Alou, Babadjou, Batcham,Batibo, Fongo Tongo, Nkongni, Santa, and Wabane
Size of geographical area (hectare)	35,000
Dominant ethnicity(ies), if appropriate	The graffi
Size of case study/project area (hectare)	1900
Dominant ethnicity in the project area	Bamilekes
Number of direct beneficiaries (people)	5000 households
Number of indirect beneficiaries (people)	1500
Geographic coordinates (latitude, longitude)	Latitudes 05 °27'N and 05 °48'N and longitudes 09°57'E and 10°15'E

finer particles from the surface horizons. The structural stability index is high, showing a stable structure (Nkembi et al. 2021).

6.1.1 Social, Cultural, Economic, and Environmental Aspects of the Study Area

The local population living within the Mt. Bamboutos ecosystem heavily depends on the agro-sylvo-pastoral sector. The average annual growth rate of the population is 2.5%. About 56% of the population is engaged in agricultural production. The Bororo minority community occupies areas of high altitudes devoted to cattle rearing (Fig. 6.4). The climate of Mt. Bamboutos is characterised by dry (November to March) and wet (April to October) seasons. Mean annual rainfall is 1918 mm, and mean temperature is 18.9 °C at an altitude of 2700 m (Kengni et al. 2009).

About 80% of the food produced is based on the exploitation of natural resources, through agriculture, livestock, forestry, hunting, beekeeping, bark harvesting, wine tapping, roots, mushrooms, termites, the collection of water, and sacrifices on sacred sites.

Data from monographs and technical reports show that agriculture and livestock farming are the main sources of income and livelihoods for the local population. The dynamism of the population based on the diversity of occupations and entrepreneurship rests on a certain number of parameters. These parameters derive from certain modes of social organisation that predispose and encourage entrepreneurship. They

Fig. 6.2 Bird's eye view of the degraded Mt. Bamboutos landscape (from Magha-Bamumbu) (Photo credit: ERuDeF)

include: traditional micro-credit, self-help groups, age groups, and individual savings in local microfinance institutions, associations, and development committees.

Before the 1970s, the local population lived in the foothill zone of Mt. Bamboutos and reserved the mountain for livestock grazing. With the decrease in soil fertility in the foothills, they migrated to the mountains to carry out both agriculture and livestock rearing. The steepness of the slopes and relatively high altitude make the land ecologically fragile. Erosion and overgrazing are leading to landscape instability, and environmental threats are intensified by poor farming practices (Ngoufo 1992).

Fig. 6.3 Agricultural practices in Mt. Bamboutos landscape (Photo credit: ERuDeF)

Fig. 6.4 The high altitudes of the Mt. Bamboutos landscape (Photo credit: ERuDeF)

6.1.2 *Farmland Management Strategies in the Mount Bamboutos Landscape*

Farmland or agricultural land is prominent among human-influenced landscapes and ecosystems (Kanianska 2016). Its proper management is very important for food security and biodiversity conservation. Given that increasing land productivity represents a key element of any equation to boost rural socio-economic growth, land degradation is increasingly viewed by local stakeholders as important, and seen as both a cause and a consequence of the perpetuation of pervasive poverty. Farmland degradation is a global phenomenon that affects human societies at the local level, where rural communities closely related to land resources are vulnerable (Amenu and Birhanu 2018). The problem of farmland degradation, which results in low agricultural productivity, is particularly severe in rural communities (Qiu et al. 2017). The main outcome of land degradation is reduction in agricultural productivity. Cropland productivity loss at a rate of 0.5 to 1% per year was estimated by Pimentel and Burgess (2013), suggesting a productivity loss of at least 20% over the last 40 years compared with a situation without soil degradation. The combined effects of continuous tillage, soil erosion, overgrazing (Frankl et al. 2011), and other factors such as cultivation of marginal lands, unsustainable use of natural resources, deforestation, and unprecedented growth of human and livestock population (Million and Belay 2004), contribute significantly to farmland degradation. Farmland degradation also leads to loss of ecosystem services. The big challenge is to balance the need to provide enough food for a growing population while at the same time maintaining healthy ecosystems and habitats (Foresight 2011).

Given the importance of ecosystem services to the sustainability and security of agricultural systems, as well as the current rate at which those services are being degraded by agricultural systems, appropriate farmland management strategies are needed (Thorn et al. 2015). Alternative practices to conventional or intensive agriculture that preserve or enhance ecosystem services without compromising farm production have been proposed and may be adopted before, during, or after cultivation (Mupangwa et al. 2012). These strategies could be active, such as surface crop residue management, or passive, such as the existence of native vegetation patches in fields. Farmland management practices may incorporate principles, among others, of multifunctional agriculture (producing food and non-food commodities, maintaining wild crop varieties, traditional landraces, and local culture (Leakey 2012), and sustainable intensification, i.e. relieving pressure on land expansion and limiting forest encroachment (Dile et al. 2013)), and conservation agriculture, or practices of no-tillage, permanent soil cover using crop residues or cover crops, and crop rotation (Friedrich et al. 2009). Such practices often require minimal inputs with opportunities for enhancing smallholder production (Mupangwa et al. 2012). Details on the effects of various strategies on farmland quality were summarised by Thorn et al. (2015). Growing leguminous cover crops fixes nitrogen, retains moisture, stimulates root growth, and encourages belowground microbial activity. No- or minimum-till systems and crop rotation influence soil organic carbon sequestration and crop yield. Fallowing helps to suppress leaching and erosion of organic matter and nutrients, and increases soil cation exchange capacity. Intercropping regulates detrimental pest populations and enhances natural enemy populations. The intercropping of trees with shade-tolerant crops, or multistory cropping, reduces the presence of weeds, and promotes nutrient cycling.

Organic and inorganic fertilisers are used for soil fertility enhancement in farmlands. Fertiliser use, particularly nitrogen (N), is an important management practice to increase crop production and improve soil fertility. Thus, the use of soil fertility-enhancing amendments to supply essential nutrients in crop production is of critical importance. Along with the nutrient supply from soil organic matter, crop residues, wet and dry deposition, and biological nitrogen fixation, synthetic (inorganic) fertilisers are also sources of essential nutrients in crop production (Jeetendra et al. 2021). Some researchers have encouraged farmers to adopt the combined application of manure and fertilisers to decrease the dependence on inorganic fertilisers (Kifayatullah et al. 2020).

The overall objective of the study was to identify and document farmland management strategies used by smallholders in the Mt. Bamboutos landscape. The specific objectives were:

1. To document the socio-demographic situation of smallholders in the Mt. Bamboutos landscape.
2. To establish the farmland soil fertility management practices used by smallholder farmers in the Mt. Bamboutos landscape.
3. To determine uptake of soil management by smallholders.
4. To determine food and nutrition security status of households within the landscape.

6.2   Methods

6.2.1   Participatory Monitoring of Impacts of the Project on the Landscape

A survey was carried out after 3 years of implementation of the project whereby farmers in this landscape were trained on soil fertility management practices, and questionnaires were administered to 203 farmers who were active in the project and willing to participate in the survey. The survey used semi-structured questionnaires to collect information on the following:

1. Socio-demographic situation of respondents
2. Uptake of soil management practices
3. Household food security and nutrition status

The questionnaires were pre-tested and corrected based on the test results before being used for data collection. Family heads were the main respondents during questionnaire administration. However, any elder person responded in the case where the family head was not available. Family heads were targeted because they make most of the decisions in the household, including those on farming. The food security situation of the population was also assessed using the household food

Fig. 6.5 Field interviews (Photo credit: ERuDeF)

insecurity access scale (HFIAS) (Coates et al. 2007). Typical scenes during questionnaire administration are presented in Fig. 6.5.

At least 20 respondents were interviewed per village. These respondents could be youths, men, or women, provided they were family heads. One focus group discussion (FGD) was conducted in each of the 20 villages, and after discussing, the interviewers assisted those who could not write, while questionnaires were handed to those who could read and write to fill in.

6.2.2 Survey Data Analysis

Data collected was processed and keyed into the SPSS software package version 6.0. Descriptive statistics (frequency, percentages), and the household food insecurity access scale (HFIAS) were used to analyse the data. Following the HFIAS, households were categorised into: severely, moderately, and mildly food insecure and food secure households.

6.3 Results

6.3.1 Socio-Demographic Situation of Respondents

The households were male dominated with 75% of respondents being male (Table 6.2). Most respondents were middle-aged people (31–50 years). People belonging to this age group are known to be at their most productive period in life. It was interesting to note that the respondents had at least a primary level of education—none of the respondents were illiterate. Large household sizes are a source of labour for family farms (Mbu et al. 2019). Children work with their parents

Table 6.2 Socio-demographic characteristics of the respondents

Variable	Description/range	Frequency	Percentage
Sex of household head	Male	152	75
	Female	51	25
Age	20–30	35	17
	31–50	68	34
	51–60	55	27
	>61	45	22
Level of education	Primary school	98	48
	Secondary school	74	36
	High school	21	10
	Higher National Diploma (HND)	6	3
	Degree holder	4	2
Duration in village (years)	<5 years	16	8
	6–10 years	3	1.5
	11–15 years	10	5
	16–20 years	5	2.5
	21 years and above	169	83
Household size	1–3	35	17
	4–6	59	29
	7–10	78	38
	11 and above	31	15
Main sources of income	Crop production	109	54
	Livestock rearing	45	22
	Others	49	24

Source: ERuDeF survey data

on the farms, especially during weekends and holidays. For most of the villagers, hiring labour was not viable, so they depended mostly on their families for labour. Large household sizes also reflect the fact that polygamy is practiced in most of the villages. Some villagers have more confidence in their family members than hired labour; the job is properly done and there are no incidents of theft. The primary source of income for the respondents was crop production, followed by animal production (Table 6.2).

6.3.2 Farmland Soil Fertility Management Practices

Farmers were trained in farm management practices and supervised through these practices during the course of the project. Farmers were able to discern the fertility status of their farmlands and determine some causes of soil fertility decline (Fig. 6.6). This was encouraging in that when farmers come to know the causes of decline, they can then make efforts to avoid or mitigate them. The results showed

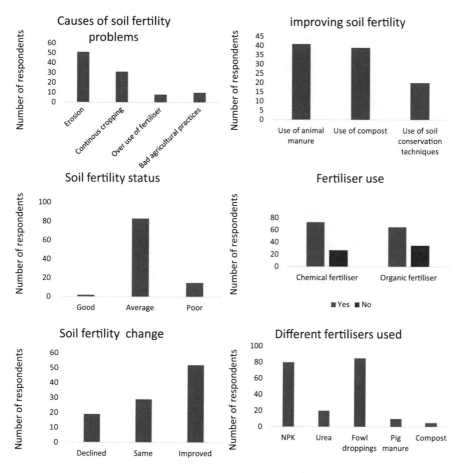

Fig. 6.6 Farmland soil fertility status and management (Source: ERuDeF survey data)

that 80% of the farmers rated their farmlands as averagely fertile (Fig. 6.6). Four major causes of soil fertility decline were evoked by the farmers, namely: soil erosion, continuous cropping, overuse of synthetic fertiliser, and bad agricultural practices (e.g. slash and burn). Erosion was identified as the main cause of soil infertility. This is logical because farmers in this area farm on steep slopes, which are easily eroded. Soil erosion reduces the agricultural value of lands via physicochemical degradation. Soil nutrient loss through runoff and sedimentation is a major driver of soil fertility decline. The eroded sediments or soil are highly concentrated with crop nutrients, which are washed away from farmlands (Bashagaluke et al. 2018).

Half of the respondent farmers acknowledged that the fertility of their soil improved, while 30% reported no change in soil fertility. Another 20% said the fertility of their farmland declined. Improved soil fertility should be linked to proper practice of lessons learnt during the project. Soil fertility improvement is visible for short-duration crops like vegetables and maize. Usage of manures improves soil fertility within a short timeframe because the nutrients from the compost are readily available for plant uptake.

The farmers used both organic and inorganic fertilisers for crop production. The main inorganic fertilisers used were nitrogen, phosphorus, and potassium, or NPK 20-10-10, and urea, while the organic fertilisers (manures) were fowl droppings, pig manure, and compost manure. Potassium (K), besides N and P, is an important nutrient needed for plant growth and development. Likewise, these three elements improve seed germination, root development, crop yield, and resistance to pests and diseases. Since most smallholder farmers do not test their soils before fertiliser application, they prefer to use compound fertiliser like NPK 20-10-10 to guarantee that all nutrients will be available for their crops. This explains why fewer farmers used urea, which is a simple fertiliser. Fowl droppings were the most used organic manure, which could be attributed to the fact that most of the farmers rear chickens in cages where they can conveniently collect the droppings for farmland fertilisation. Some buy fowl droppings from the nearby city of Bafoussam, where there are large poultry farms.

The farmers reported the use of animal manure, compost manure, and good farming practices as ideal means of improving soil fertility (Fig. 6.6). Manure is a natural fertiliser containing essential elements required for plant growth. Its application to cropland restores or replenishes soil fertility. It reduces soil erosion, restores eroded croplands, and reduces nutrient leaching. All these factors positively affect crop yield, and are therefore expected to contribute to food security (Teenstra et al. 2016).

6.3.3 Cropping Systems Practiced

Several cropping systems were practiced in production landscapes as a means of maximising land productivity. The results showed that access to land greatly determined the cropping system practiced by farmers (Fig. 6.7). Farmland cultivated was either owned or rented, and most farmers practiced restricted types of cropping systems on rented lands. On their own land, they used a variety of cropping systems. The restriction could be attributed to the different terms of agreement for those who lease land. Therefore, the mode of access to land influences the management of farmlands in the landscape. Live fencing and alley cropping were not reported on leased land. A detailed study on buying and renting land in the Bamboutos mountains by Nzoffou et al. (2020) found that two types of rental systems were practiced: long term and short term. A long-term lease referred to a contract that had at least a 5-year renewable term with a royalty paid per crop year. A short-term rental referred to contracts with a renewable one-year term. In short-term rentals, the tenant is called upon to engage in new negotiations at the end of the contract, and the full rental fee is generally paid before the start of the crop year. The tenant must not plant perennial

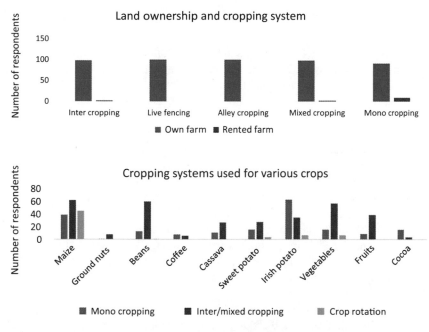

Fig. 6.7 Cropping systems and crops planted (Source: ERuDeF survey data)

Fig. 6.8 Sample mixed cropping system (Photo credit: ERuDeF)

crops and trees, only short-cycle crops are allowed. The results of this study show that leased farmlands were on a short-term basis.

The farmers had the choice of what crop to use for which cropping system. Cash crops (cocoa and coffee) were planted mainly in mono and mixed cropping systems. Short-cycle crops (maize, beans, Irish potatoes), in addition to being planted in mono and mixed cropping systems, were also planted in crop rotational systems. In a country-wide study, Abia et al. (2016) reported that mixed cropping or crop rotation could go a long way to solve problems faced and boost future crop production in Cameroon. Mixed cropping or crop rotation helps to reduce the spread of rapid attacks on crops by pests and diseases, hence increasing food availability and supply. Sample mixed cropping systems are presented in Fig. 6.8.

6.3.4 Agroforestry Practices and Organic Manure Production

The agroforestry system in this area refers to scattered trees on cropland—an agrosilvicultural agroforestry system. In this system, trees and shrubs are either planted or maintained within agricultural lands (Njukeng et al. 2021) (Fig. 6.9). They could equally be described as homegardens. During implementation of the project, emphasis was placed on agroforestry practices because of their importance for ecosystem restoration and farm diversification. Farmers were trained and supervised on the different components of agroforestry practices and organic manure production methods. The farmers gave four main reasons for practicing agroforestry: improved crop yields, fuel wood, fodder, and diverse income. The main components of agroforestry practiced by farmers in the study site were: planting of fertilsser, fodder, fruit, and timber trees in cropland. The main ways of producing organic manure were through mulching, composting, and farmyard manure preparation (Fig. 6.10). The three main constraints for practicing agroforestry and preparing organic manures were stated to be: lack of training, lack of seeds, and excessive labour. Examples of fertiliser and fruit tree-based agroforestry systems are presented in Fig. 6.10.

Fertiliser trees are known to add nutrients to the soil through nitrogen fixing in the nodules or when their leaves fall and are incorporated into the soil. The main species planted in the study area was *Leucaena leucocephala*, which is used as manure and fodder. Fodder trees are used to feed animals. Their leaves and soft branches are harvested and fed to animals. Fruit trees provide food for family consumption as well as for income generation. The most planted fruit tree in the study area was the avocado (*Persea americana*). Timber trees are used for construction, furniture production, or sold for income generation. They also provide fuel wood for farmers.

Fig. 6.9 Fertiliser and fruit trees in agroforestry systems (Photo credits: ERuDeF)

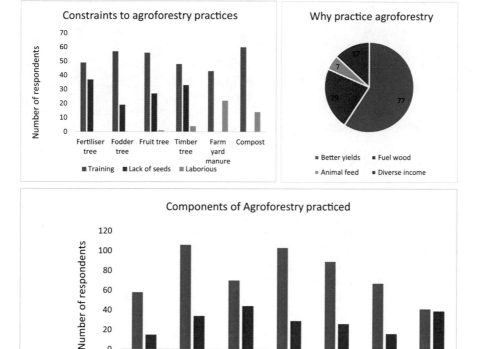

Fig. 6.10 Agroforestry and organic manure production practices and constraints (Source: ERuDeF survey data)

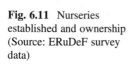

Fig. 6.11 Nurseries established and ownership (Source: ERuDeF survey data)

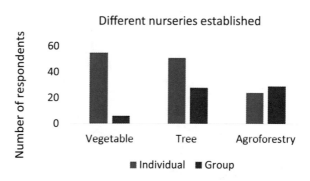

6.3.5 Nursery Establishment

Nurseries were established either as community (group) or individual nurseries. Vegetable, tree, and agroforestry nurseries were established (Fig. 6.11). Most of

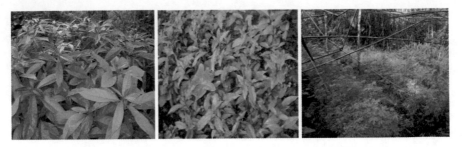

Fig. 6.12 Group nurseries of fruit trees (Avocado (*Persea americana*), timber trees, (*Entandrophragma angolense,*), and fertiliser/fodder trees (*Leucaena leucocephala*) (from left to right) (Photo credits: ERuDeF)

the vegetable nurseries were owned by individuals, while agroforestry and tree nurseries were mostly owned by community groups. Tree nurseries included fruit trees and timber trees, while agroforestry nurseries contained fodder and fertiliser tree species. Group nurseries (Fig. 6.12) were generally larger than individual nurseries. This is because there are many community members belonging to the group that dedicates time to take care of the nurseries, compared to individual nurseries where labour is limited to the nursery owner and family.

6.3.6   Food and Nutrition Security

Soil fertility management on farmland was undertaken to ensure good and sustainable crop yields and hence food and nutrition security. Farmers plant different crops for diversified income and nutrition. The main staple crops were maize, Irish potatoes, and plantain (Fig. 6.13). Of all respondents, 49% acknowledged that their food production had improved, while 51% responded that they did not realise

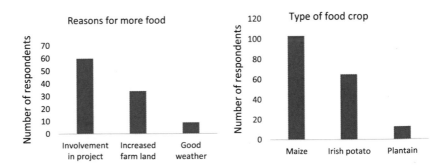

Fig. 6.13 Staple foods and reasons for increased food production in the study area (Source: ERuDeF survey data)

Fig. 6.14 Households food security status in the study area (Source: ERuDeF survey data)

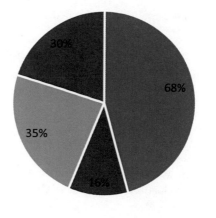

- Food secured
- Mildly food insecure
- Moderately food insecure
- Severely food insecure

any improvement. The research team analysed the reasons for food improvement and found that involvement in the project was the primary reason, followed by an increase in farmland, and lastly, good weather conditions (Fig. 6.13).

The food security situation of the studied population was assessed. Results showed that 35% of respondent households were food secure, and 30% were severely food insecure (Fig. 6.14). Efforts to address food security in the landscape should be intensified to increase the number of food secure households.

6.4 Conclusion and Lessons Learnt

Farmers have acquired numerous skills in farmland soil fertility management. However, practice of any of the techniques was highly dependent on land access. The farmers used both organic and inorganic (synthetic) fertilisers for crop production as well as agroforestry practices. Acquiring these skills has gone a long way towards improving food and nutrition security at the household level. Constraints in practicing agroforestry and organic manure production were inadequate training, lack of seeds, ignorance, and the amount of labour required for some of the activities. Lack of information about good farm management practices was evident. This ignorance is an indication that training and sensitisation on good soil fertility management practices need to continue. Continuation would help to further improve the food and nutrition situation of households. Likewise, proper farmland soil management would reduce encroachment into the mountain landscape and thus conserve biodiversity.

The future challenges identified during the course of this study are:

- Sustaining coordination and communication between the different stakeholders.
- Perverse incentives that promote unsustainable practices (e.g. excessive use of chemicals in farming).
- Effects of climate change and extreme weather conditions.

Lessons Learnt
- Good farmland management practices led to improved farm yields, thus improving the situation of food security.
- Agroforestry and mixed cropping practices helped farmers to diversify their farm produce as well as income sources. This helped the farmers to rely mostly on their farm production rather than relying on illegal harvesting of products from the forest.
- Capacity building should be enhanced to ensure that a good number of farmers master the various soil fertility management practices. This will ensure better crop yields to ensure food security and reduce reliance on the forest, thus conserving biodiversity.

Acknowledgements We gratefully acknowledge the financial support of TreeSisters and the Darwin Initiative, and the technical support of the International Tree Foundation (ITF) and the International Union for the Conservation of Nature (IUCN). The collaboration of local communities, especially the smallholder farmers, is highly appreciated.

References

Abia WA, Shum CE, Fomboh RN, Ntungwe EN, Ageh MT (2016) Agriculture in Cameroon: proposed strategies to sustain productivity. Int J Res Agric Res 2(2):1–3

Abubakar AS, Fondo TA, Nyong PA (2020) Agroforestry for sustainable agriculture in the Western highlands of Cameroon Haya. Saudi J Life Sci 5(9):160–164

Amenu L, Birhanu I (2018) Determinants of farmland degradation and its implication on crop productivity and sustainability. J Appl Sci Environ Manag 22(8):1225–1232

Andrianarison F, Housseini B, Oldiges C (2022) Dynamics and Determinants of monetary and multidimensional poverty in Cameroon. OPHI Working Paper 141. University of Oxford

Bashagaluke JB, Logah V, Opoku A, Sarkodie-Addo J, Quansah C (2018) Soil nutrient loss through erosion: impact of different cropping systems and soil amendments in Ghana. PLoS One 13(12):e0208250

Burke K (2001) Origin of the Cameroon Line of Volcano-Capped Swells. J Geol 109:349–362. https://doi.org/10.1086/319977

Coates J, Swindale A, Bilinsky P (2007) Household food insecurity access scale (HFIAS) for measurement of household food access: indicator guide, vol 3. FHI 360/FANTA, Washington, D.C.

Dile YT, Karlberg L, Temesgen M, Rockström J (2013) The role of water harvesting to achieve sustainable agricultural intensification and resilience against water related shocks in sub-Saharan Africa. Agric Ecosyst Environ 181:69–79

Ewane BE, Asabaimbi DN, Njiaghait YM, Nkembi L (2021) Agricultural expansion and land use land cover changes in the mount Bamboutos landscape, Western Cameroon: implications for local land use planning and sustainable development. Int J Environ Stud 80:186. https://doi.org/10.1080/00207233.2021.2006911

Frankl A, Nyssen J, De Dapper M, Mitiku H, Deckers J, Poesen J (2011) Trends in gully erosion as evidenced from repeat photography (North Ethiopia). Landform Analys 17:47–50

Friedrich T, Kienzle J, Kassam A (2009) Conservation agriculture in developing countries: the role of mechanization. In: Seminar on innovation for sustainable agricultural mechanisation, Hannover, Germany, 8 November 2009. Food and Agricultural Organization

Foresight (2011) The future of food and farming executive summary. The Government Office for Science, London

Jeetendra PA, Sapkota TB, Krupnik TJ, Rahut DB, Jat ML, Stirling CM (2021) Factors affecting farmers' use of organic and inorganic fertilizers in South Asia. Environ Sci Pollut Res 28: 51480–51496

Kanianska R (2016) Agriculture and its impact on land-use, environment, and ecosystem services. In: Almusaed A (ed) Landscape ecology. Intech Open

Kengni L, Tekoudjou H, Tematio P, Pamo Tedonkeng E, Tankou CM, Lucas Y, Probst JL (2009) Rainfall variability along the southern flank of the Bambouto mountain (West-Cameroon). J Cameroon Acad Sci 8(1):45–52

Kifayatullah K, Xuan TD, Zubair N, Shafiqullah A, Gulbuddin G (2020) Effects of organic and inorganic fertilizer application on growth, yield, and grain quality of rice. Agriculture 10:544

Leakey RB (2012) Multifunctional agriculture and opportunities for agroforestry: implications of IAASTD. In: Nair PKR, Garrity D (eds) Agroforestry-the future of global land use. Dordrecht, Springer, pp 203–214

Mbu DT, Nganje SN, Chuo JN (2019) Intricacies of organic and chemical fertilizer application on arable land crop production in Cameroon. J Socioecon Develop 2(2):61–72

Million T, Belay K (2004) Factors influencing adoption of soil conservation measures in Southern Ethiopia: the case of Gununo area. J Agric Rural Develop 105(1):49–62

Mupangwa W, Twomlow S, Walker S (2012) Reduced tillage, mulching and rotational effects on maize (Zea mays L.), cowpea (Vigna unguiculata (Walp)L.) and sorghum (Sorghum bicolor L. (Moench)) yields under semi-arid conditions. Field Crop Resarch 132:139–148

Ngoufo R (1992) The Bamboutos mountains: environment and rural land use in West Cameroon. Mt Res Dev 12(4):349–356

Nkembi L, Deh NH, Tankou CM, Njukeng JN (2021) Analysis of small scale farmers households food security in the mount Bamboutos ecosystem. J Food Secur 9(2):56–61. https://doi.org/10.12691/jfs-9-2-3

Njukeng JN, Nkembi L, Tankou CM, Deh NH, Ngulefack EF (2021) Soil fertility management practices by smallholder farmers in the Bamboutos Mountain ecosystem. World J Agric Res 9(2):58–64. https://doi.org/10.12691/wjar-9-2-3

Nzoffou JL, Fongang GH, Kaffo C (2020) Buying and land rental in the mountains: what are the challenges for the development of modern farms? Case of the Bamboutos Mountains - West - Cameroon. Int J Rural Develop Environ Health Res (IJREH) 4(4):154–162

Pimentel D, Burgess M (2013) Soil erosion threatens food production. Agriculture 3(3):443–463

Qiu LL, Koondhar MA, Liu Y, Zeng WZ (2017) Land degradtion is the instinctive source of poverty in rural areas of Pakistan. IOP Conf Series Earth Environ Sci 86(1):012003

Tankou CM, de Snoo GR, de Iongh HH, Persoon G (2013) Soil quality assessment of cropping systems in Western highlands of Cameroon. Int J Agric Res 8(1):1–16

Teenstra E, Andeweg K, Vellinga T (2016) Manure helps feed the world: integrated manure management demonstrates manure is a valuable resource. In: Climate-smart agriculture practice brief. CGIAR Research Program on Climate Change, Agriculture and Food Security (CCAFS), Copenhagen

Tematio P, Kengni L, Bitom D, Hodson M, Fopoussi JC, Leumbe O, Mpakam HG, Tsozué D (2004) Soils and their distribution on Bambouto volcanic mountain, West Cameroon highland, Central Africa. J Afr Earth Sci 39:447–457

Thorn J, Snaddon J, Waldron A, Kok K, Zhou W, Bhagwat S, Willis K, Petrokofsky G (2015) How effective are on-farm conservation land management strategies for preserving ecosystem services in developing countries? A systematic map protocol. Environ Evidence 4(11):1–13

Toh FA, Angwafo T, Ndam LM, Antoine MZ (2018) The socio-economic impact of land use and land cover change on the inhabitants of mount Bamboutos caldera of the Western highlands of Cameroon. Adv Remote Sensing 7:25–45

Yerima BPK, Van Ranst E (2005) Major soil classification systems used in the tropics: soils of Cameroon. Trafford Publishing. 1-4120-5789-2

The opinions expressed in this chapter are those of the author(s) and do not necessarily reflect the views of UNU-IAS, its Board of Directors, or the countries they represent.

Open Access This chapter is licenced under the terms of the Creative Commons Attribution-NonCommercial-ShareAlike 3.0 IGO licence (http://creativecommons.org/licenses/by-nc-sa/3.0/igo/), which permits any noncommercial use, sharing, adaptation, distribution and reproduction in any medium or format, as long as you give appropriate credit to UNU-IAS, provide a link to the Creative Commons licence and indicate if changes were made. If you remix, transform, or build upon this book or a part thereof, you must distribute your contributions under the same licence as the original. The use of the UNU-IAS name and logo, shall be subject to a separate written licence agreement between UNU-IAS and the user and is not authorised as part of this CC BY-NC-SA 3.0 IGO licence. Note that the link provided above includes additional terms and conditions of the licence.

The images or other third party material in this chapter are included in the chapter's Creative Commons licence, unless indicated otherwise in a credit line to the material. If material is not included in the chapter's Creative Commons licence and your intended use is not permitted by statutory regulation or exceeds the permitted use, you will need to obtain permission directly from the copyright holder.

Chapter 7
Traditional Regenerative Agriculture as a Sustainable Landscape Approach: Lessons from India and Thailand

Yoji Natori, Pia Sethi, Prasert Trakansuphakon, and Siddharth Edake

7.1 Introduction

Systems of agriculture such as shifting or rotational cultivation involving cycles of brief periods of crop cultivation and prolonged fallows play a key role in harmonising farming and conservation, and also contribute to ecosystem restoration. These systems use the functions of the forest during fallow periods to replenish soil fertility,[1] allowing crops to be cultivated without external inputs of fertilisers. Proximity of fields to primary or late-successional forests facilitates quicker regeneration back to forest through seed dispersal and pollination. Enhanced pollination services also boost agricultural productivity. Fields are prepared for cultivation by burning plant residues. Farmers practice polyculture for one to two seasons, after which they leave the plot fallow and "shift" to a new patch (thus, this practice is often called "shifting cultivation", or alternatively, "rotational farming", as farmers come

---

[1] Shifting cultivation or swidden agriculture is practiced in the nutrient-poor soils of tropical uplands. Rather than depend on the use of external inputs, the fallow phase is used instead to facilate nutrient uptake and accumulate biomass, which returns to the soil as litter. The nutrients that accumulate in the biomass become available for the next crop when the fallow vegetation is felled and burned.

Y. Natori (✉)
Akita International University, Akita, Japan
e-mail: ynatori@aiu.ac.jp

P. Sethi
Centre for Ecology, Development, and Research (CEDAR), Dehradun, Uttarakhand, India

P. Trakansuphakon
Pgakenyaw Association for Sustainable Development (PASD), Chiang Mai, Thailand

S. Edake
The Energy and Resources Institute (TERI), New Delhi, India

© The Author(s) 2023   117
M. Nishi, S. M. Subramanian (eds.), *Ecosystem Restoration through Managing Socio-Ecological Production Landscapes and Seascapes (SEPLS)*, Satoyama Initiative Thematic Review, https://doi.org/10.1007/978-981-99-1292-6_7

Fig. 7.1 Habitat mosaics created by traditional regenerative agriculture (Source: Authors' creation)

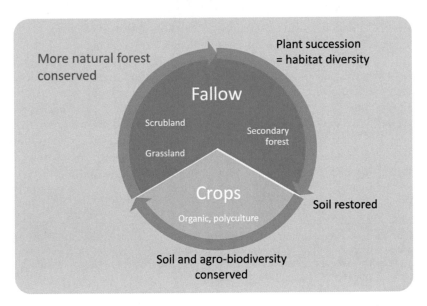

Fig. 7.2 Traditional regenerative agriculture that supports agriculture and biodiversity conservation at landscape level (Source: Authors' creation)

back to the plot in rotation). This traditional form of agriculture regenerates nature and forests in the process by drawing upon nature's regenerative potential; hence, we have termed it "traditional regenerative agriculture" (TRA). Because it uses plant succession to regenerate soil fertility, TRA creates a mosaic of habitat types at the landscape level, contributing to biodiversity (Fig. 7.1). At the same time, since farmers repeatedly use the same areas in the cycle, it leaves other areas to natural processes (Fig. 7.2). In South and Southeast Asia, shifting cultivation is also often a part of a larger system including uncut forests, settled paddy farming along water sources, home gardens, and tree cultivation (e.g. Cairns and Brookfield 2011).

Because the cycle requires clearing and burning of forest, this practice is nevertheless often considered to be destructive and environmentally harmful. Likewise, the fallow period—the key segment of the cycle that makes TRA ecologically sound—is often viewed as wasteful and inefficient (e.g. Borah and Goswami 1973; Ranjan and Upadhyay 1999). Communities face pressure to adopt settled forms of agriculture, although these may be harmful to the environment and livelihoods of people engaged in TRA (Pant et al. 2018), and inappropriate for the local ecology, culture (e.g. Teegalapalli and Datta 2016), and climate.

Here, we critically examine the evidence in support of TRA to highlight its value as a time tested, ecosystem restoration strategy. For this purpose, we describe below: (1) how TRA functions as a form of ecosystem restoration; (2) the context that has led to the shortening of fallow periods, with several resultant negative environmental consequences; and (3) the practices that address the shortcomings of these short fallows in Northeast India and Northern Thailand. Based on empirical evidence, we make a case for TRA's continued practice. A diversity of land uses potentially enhances ecosystem resilience and creates a diversified portfolio of responses in the face of climate and ecosystem change. Recognising the significance of TRA will help to sustain resilient, sustainable, and culturally appropriate socio-ecological production landscapes in some of the most biodiverse regions of the planet.

7.2 TRA as a Form of Ecosystem Restoration and Factors that Threaten It

Contrary to popular misconceptions, areas under TRA help the world to mitigate climate change (Bruun et al. 2009) whilst meeting crucial food demands and providing a diversity of habitats for biodiversity conservation (Finegan and Nasi 2004; Rerkasem et al. 2009). Our ecological knowledge allows us to develop the following model, which illustrates ecological succession using carbon stock as a surrogate, in shifting cultivation cycles under different fallow periods (Fig. 7.3). Plants and trees sequester carbon dioxide and store it in the ecosystem at varying rates depending on the age of fallows (Gogoi et al. 2020). In the optimal TRA cycle (represented by the blue curve in Fig. 7.3; Type A in Table 7.1), the carbon stock in the ecosystem returns to pre-existing levels at the end of the fallow period. As farmers cultivate different plots, they create landscape mosaics of different successional stages (Fig. 7.1). The longer the TRA cycle, the more variation in successional stages there will be in the TRA system, which is expected to support higher biodiversity. That is, at any given time, there will be plots of land located along the curve, representing habitat diversity resulting from different stages of ecological succession within the landscape. TRA therefore actively contributes to ecosystem

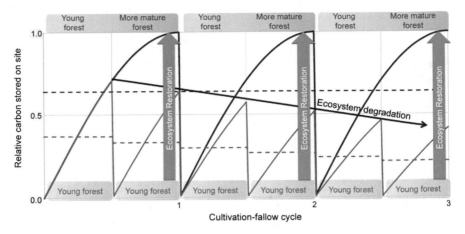

Fig. 7.3 Model showing the environmental consequences of different lengths of the fallow period (Source: Authors' creation). The x-axis represents the number of optimal cycles, the length of which is highly site dependent. An optimal cycle could be 7–12 years in Northern Thailand and 15–20 years in Nagaland. Carbon storage is approximated with sine curves. The dotted lines show the time-averaged carbon stock of the curves of the same colour over the cultivation-fallow cycle

restoration and enables conservation at the landscape level. This dual value of TRA for both biodiversity conservation and food production is rarely recognised.

When the TRA cycle is shortened (orange curve in Fig.7.3; Type B in Table 7.1), the average carbon stock across the TRA cycle (horizontal lines) declines. This occurs due to the shorter time available for the fallow plots to sequester carbon between cultivations, unless active measures are taken to accelerate carbon sequestration. The difference between the average carbon stocks (i.e. difference between the blue and orange horizontal lines) represents net carbon emissions resulting from the reduced cycles. In addition, the shorter fallow periods result in reduced productivity due to insufficient site recovery, which causes progressive reductions in carbon stock in subsequent cycles—this is expressed in the model by a 10% reduction in carbon stock recovery in each cycle.

An empirical study in Southeast Asia (Bruun et al. 2009) indicated that time-averaged aboveground carbon stocks decline by about 90% when fallow periods are reduced to 4 years, and by about 60% if fallows are converted to oil palm plantations. Habitat diversity, too, is reduced under a shortened TRA cycle, due to reduced time for ecosystems to recover. Under these conditions, the default is adoption of more intensive land use of short fallows (i.e. Type B) or conversion to permanent farms. Type B could further result in the conversion of currently untouched forests to farmland to compensate for the reduced yield (due to the decreasing soil quality and reduced carbon stock of shorter cycles; Bruun et al. 2009).

Changing farming practices alter the nature of TRA, which creates modified suites of benefits and shortcomings (Table 7.1). The benefits of long fallow periods—the traditional practice—are mainly environmental, whilst the

Table 7.1 Benefits and shortcomings of long and short fallow periods (compiled from the authors' observations)

	Type A: Long allow TRA (traditional practice)	Type B: Short fallow TRA (prevailing trend in NE India and Thailand)
Benefits	• Replenishes soil fertility and reduces soil erosion • Maintains consistent yields • Provides diverse habitat types; high biodiversity • Regenerates back to secondary forest • Enhances probability of continued use of traditional crop varieties	• Addresses the inefficient-use criticism • Addresses issues of labour shortage • Increases the probability of farms being closer to habitation; decreased opportunity costs
Shortcomings	• Criticism from the wasteful use perspective • Risk of being converted to settled agriculture or "productive" plantations (a consequence of the low productivity stereotyping). • Labour shortages make it difficult to continue • Increased distances to "*jhumed*" land; more time demanding	• Insufficient soil recovery; low soil fertility and high erosion • Diminishing yields, increasing food insecurity • Simplified habitat assembly; reducing biodiversity • Decreased agro- and plant species diversity; increased probability of replacement of traditional disease- and drought-resistant crop varieties with commercial crops and/or plantations
Ways to address shortcomings	• Value addition to non-timber forest products (NTFPs) • Supplement with agroforestry plantations • Strengthen market linkages, product development, and value addition • Appropriate policy and legal frameworks that recognise TRA as a legitimate form of agriculture[a]	• Rapid forest recovery measures for habitat creation and soil fertility replenishment • Measures to reduce soil erosion • Research and capacity building to ensure proper practices • Application of traditional knowledge/practices to enhance crop productivity and fallow recovery

[a]This is critically important for various reasons and applies to Type B as well. In India, because of the communal nature of shifting cultivation, farmers cannot access credit from banks and other financial agencies as they lack titles to the land they cultivate. In Thailand, the ambiguity of rotational farming, falling between forestry and agriculture, makes it difficult for rotational farmers to have access to agricultural support from the government

shortcomings are mainly social and economic. Shortening the fallow period, which is a common way of addressing these shortcomings, potentially leads to ecosystem degradation. Below, we describe traditional practices that can help avoid the unproductive consequences of a short fallow period by aiding forest recovery.

7.3 Traditional Techniques for Rapid Forest Recovery in Short Fallow Cultivation Cycles

The shortening of shifting cultivation cycles is increasingly the norm in Northeast India and Northern Thailand. Though reasons vary from one site to another (see below), there is a shared concern for the negative consequences of deviating from traditions. At the same time, both countries face pressure to adopt alternative, permanent cropping systems (Maithani 2005) and/or to convert to oil palm (Srinivasan 2014) and other tree plantations. Here, we review two cases where indigenous communities use their traditional knowledge to speed up forest recovery after cultivation. These two cases may provide valuable insights for other sites facing similar challenges.

7.3.1 Nagaland, India

Nagaland is a state in Northeast India (Fig. 7.4, Table 7.2). It is located in one of the most biodiverse regions of the world, where the Himalayas and Indo-Burma Biodiversity Hotspots meet. In Nagaland, high forest cover coexists with shifting cultivation (*jhum*). Although the Indian government has been promoting sedentary farming, particularly since *jhum* is viewed as a major cause of deforestation in Northeast India, *jhum* persists as the farming practice most suited to the topography and cultural sensitivities of the region. However, the length of the fallow period has

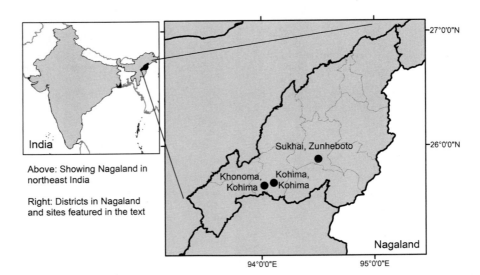

Fig. 7.4 Case study sites in Nagaland in Northeast India (Source: Authors' creation based on district boundaries by Meena (2018), on the background of Natural Earth free map data at naturalearthdata.com)

Table 7.2 Basic information of the study area, Nagaland, India

Country	India
Province	Nagaland
District	Kohima (featured case)
Municipality	Khonoma Village (featured case)
Size of geographical area (hectare)	1,657,900 (State of Nagaland)
Dominant ethnicity(ies), if appropriate	The Naga tribes
Size of case study/project area (hectare)	n/a
Dominant ethnicity in the project area	16 major tribes in Nagaland; Angami in Khonoma in Kohima district
Number of direct beneficiaries (people)	n/a
Number of indirect beneficiaries (people)	n/a
Geographic coordinates (latitude, longitude)	25°40'31"N, 94°06'19"E (state capital: Kohima)

(Source: Authors)

decreased in Nagaland. Conversion to oil palm plantations—in search of economic returns—is an emerging and eminent threat as conversion often comes at the expense of environmental and cultural values (e.g. Pant et al. 2018 and references therein).

The increased food demand due to population increase in villages,[2] possibly along with recent shortages of labour due to out-migration, has progressively shortened the fallow period in Nagaland from the traditional 15–20 or more years to 10–12 years in the late nineteenth century,[3] and as low as 5–10 years (average of 9) today (Hutton 1921; Ramakrishnan 1992; Cairns et al. 2007; Rathore et al. 2010; NEPED and IIRR 1999). This trend poses a threat to the environmental sustainability (Ramakrishnan, 1992; Bruun et al. 2009) of the traditional system.

---

[2] The population of Nagaland increased by 64.41% between 1991 and 2001 as compared to the all-India average of 21.34% per the Census of India in 2001, although the accuracy of some of the census figures have been questioned (Agrawal and Kumar 2012). Nagaland is the only state in India which saw an increase in area under *jhum* between 2005–2006 and 2015–2016 (NRSC & DOLR, 2005–2015).

[3] Even a hundred years ago, regional variation existed. Hutton (1921, p. 59) has this to say about the length of the cycle in what is now Zunheboto district, Nagaland: "*In the Tizü valley, however, and in parts of Kileki valley where the population has much outgrown the supply of suitable jhuming land, jhums may often be found cleared after only five years' rest, and in some villages even after three, while loads of earth have to be sometimes actually carried and dumped down in the rocky parts of the field to make sowing possible at all.*"

Fig. 7.5 Alder-based *jhum* in Khonosma village, Nagaland. Crops are cultivated between pollarded alder trees (Photo by S. Edake, January 2021)

Despite the shortcomings of shifting cultivation with short fallow periods, highly sophisticated indigenous practices exist that circumvent issues of reduced fertility or enhanced soil erosion. The Angami community members of Khonoma village in Kohima district, Nagaland, for example practice a traditional agroforestry system based on the Nepalese alder (*Alnus nepalensis*) believed to be at least 500 years old (Cairns 2007) (Fig. 7.5). The alder is a deciduous or semi-deciduous, pioneer species that grows naturally throughout the Himalayas. Symbiotic actinomycetes *Frankia* on the root nodules help this tree species to fix nitrogen and improve soil fertility.

The indigenous tribes of Nagaland likely developed the alder system of *jhum* through astute observation and natural experimentation (Kehie et al. 2017). Representatives of Khonoma village explained their traditional practices to this chapter's authors during site visits in January 2021 and August 2022.

In each shifting cultivation cycle, the Angami farmers in Khonoma village do not fell the alder trees but only prune 2–2.5 m from the ground. Crops are intercropped with the alder trees. There is no standard rule regarding spacing of alder trees, although about 6 m is common (Gokhale et al. 1985). Pollarded stumps are covered with mud/straw to prevent drying, and stone slabs are placed over them to allow the uniform sprouting of new shoots (Cairns et al. 2007). Farmers prune the alders in November and December. Amongst the resprouting stems, only the most vigorous stems are kept, with others removed the following August. Pollarding is reported to provide a silvicultural benefit of stimulating the growth of alder trees, and might even prolong their longevity (Cairn et al. 2007). Some trees in Khonoma are 150–200 years old, whilst the same tree species is considered a short-lived species (Lamichhaney 1995) in Nepal.

The area under *jhum* would be divided into five units. Each would be cultivated for 2 years and they would return to the original patch after 10 years (Meyase, pers.

comm.). In the first year, upland rice is planted in warmer areas and Job's tears (*Coix lacryma-jobi*) in cooler areas, frequently interplanted along with garlic, beans, chillies, cucurbits, and taro as secondary crops. Crops are harvested from October to November, and cattle are allowed to graze. Maize and pearl millet are grown during the second year. Following this, the lands are left fallow to regenerate from August onwards during which time they provide fuelwood, fix nitrogen, and accumulate soil organic matter (Cairns et al. 2007). The farmers may also divide their lands into 10 small plots. After a year, they shift to the next plot to cultivate, returning to the original plot after 10 years.

To prepare a field for cultivation, shrubs, and grasses are cut. Next, the alder tree branches are felled and collected for firewood and timber. Small branches and leaves are left to dry along with shrubs and grass. Once dry, the farmers burn them. The farmers also leave the dry leaves of the alder tree, which decompose and fertilise the soil. The alder trees start regrowing naturally. The farmers do not cultivate the forests on the top of the hillsides, which harbour water sources. Instead, *jhum* is practiced only on lower slopes.

The *jhum* farmers in Khonoma cultivate smaller areas, and hence labour shortage is not an issue. An important product derived from alder-based *jhum* is firewood and poles. The farmers, therefore, concentrate more on collecting firewood than cultivating agricultural products in *jhum* fields. This is also possibly due to the high productivity of wet terraced rice that is grown in the valley bottoms in addition to *jhum*. The farmers do not use machines, as they feel that the use of chainsaws could hinder the regrowth of the alder trees. Hence, they prefer to use axes to cut the branches. The *jhum* farmers grow a variety of crops like maize, cabbage, potatoes, and chillies. Potatoes are sold at a price rate of INR 50/kg (USD 0.67 on 2022/1/17). Chillies and maize are cultivated for self-consumption, but some maize is used for chicken feed. The farmers say they have not seen any decreases in productivity.

The farmers of Khonoma have an innovative way of preventing soil erosion. Their terraced *jhum* fields do not have barriers to hold the soil. Instead, the paddy fields are located right below the terraced farms, so that the rain washes the soil directly into the paddy fields below. The horizontal spreading root systems of the alder tree also help to stabilise the slopes and reduce soil erosion. In Khonoma, another N-fixing species, *Albizia stipulata*, is planted to enhance the fertility of wet rice cultivation (Cairns 2007). The Angamis of Khonoma practice a highly sophisticated, intensified system of TRA with several unique facets that are not commonly observed in other places where *jhum* is practiced.[4]

Jhum lands in Nagaland exhibit exceptionally high levels of agro-biodiversity. Chetehba village in Phek district has 167 types of crops, including 12 rice varieties. In the Wokha district, 18 residual crops remain in the fields after the main crops are

---

[4] For example, the creation of terraces for *jhum* is not a common practice in shifting cultivation, nor is the presence of stone walls. While 10-year cycles are followed possibly because of the resulting pole yield, a 4-year total cycle is reportedly effective at regenerating *jhum* fallows (see Cairns 2007).

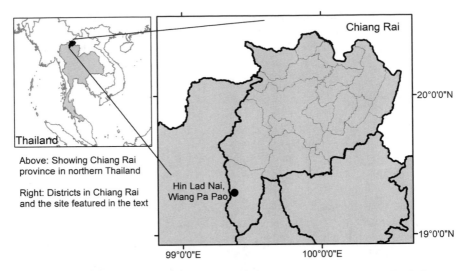

Fig. 7.6 Case study site, Hin Lad Nai community in Chiang Rai Province, Thailand (Source: Authors' creation based on the boundaries of provinces and districts by Royal Thai Survey Department (2019), on the background of Natural Earth free map data at naturalearthdata.com)

harvested (NEPED and IIRR 1999). At a three-day farmer's market in September 2021, 66 varieties of vegetables and fruits were brought for sale from nine villages in Zunheboto District (TERI, unpublished data).

7.3.2   Chiang Rai, Thailand

The Karen people of Northern Thailand refer to shifting cultivation as *Baa muan wiang*, literally "rotating forests" (Cairns et al. 2007), or *rai mun wian* ("rotational farming"). The majority view of the state, and the public, however, is that rotational farming is detrimental to forests and biodiversity, associated with opium production, and practiced by people disloyal to the Thai nation, Buddhism, and the Thai monarchy. There are many cases of rotational farmers being arrested or forced to stop the practice due to direct and indirect pressure from the national government's development agenda, and bilateral and multilateral development aid. The state encourages the growing of cash crops (such as cabbages, potatoes, and corn) in permanent farms with the use of chemical pesticides and herbicides. The consequences, however, are an increased impact on the environment, and financial hardship and health problems for farmers. For the purposes of this chapter, we highlight the traditional farming practices of the community of Hin Lad Nai in Chiang Rai Province in Northern Thailand (Fig. 7.6; Table 7.3).

There is, nevertheless, some level of acceptance in Thailand that rotational farming is a system that supports food security because a wide variety of

Table 7.3 Basic information of the study area, Hin Lad Nai, Chiang Rai, Thailand

Country	Thailand
Province	Chiang Rai
District	Wiang Pa pao
Municipality	Hin Lad Nai
Size of geographical area (hectare)	Approximately 3,700
Dominant ethnicity(ies), if appropriate	The Karen people, Lahu people
Size of case study/project area (hectare)	n/a
Dominant ethnicity in the project area	The Karen people
Number of direct beneficiaries (people)	n/a
Number of indirect beneficiaries (people)	n/a
Geographic coordinates (latitude, longitude)	19°18'12.3"N, 99°22'15.1"E

(Source: Authors, SwedBio and PASD 2016)

traditional/local seeds and plants are grown. What has proven more difficult to convey is that the fallow forest is an integral—and important—part of the cultivation cycle. The authorities consider the long fallow period as a wasteful use of land, and rotational farmers are ineligible for the public support that other farmers ordinarily enjoy. To cope with this situation, rotational farmers are innovating ways to quickly regenerate soil fertility, to shorten the fallow period, and to increase the productivity of fallow land with non-timber forest products. Fallow land regenerates more quickly with species such as *Macaranga denticulata* (or *p'dav* in the Karen language). A shorter fallow period addresses the criticism of rotational farming as a wasteful land use, and faster re-establishment of a forest mitigates the shortcomings of Type B TRA discussed above.

The *p'dav* grows in fallow forests (Fig. 7.7). The Karen people who traditionally use *p'dav* also know when the land is ready for another season of farming. A Karen poem says, "*p'dav fallow grows up in a line along fallow land, why do you keep farming the old field?*" Karen traditional knowledge contrasts two diverse ways of farming—one is opening the good fallow land for farming again, and another is continuing to cultivate the old cropland—and questions the use of unproductive old cropland when the *p'dav* trees presage the readiness of the fallow land to farm and obtain good yields. The *p'dav* is an indicator of the readiness of the fallow site for re-cultivation. Traditional rotational farming embodies a living culture where such invaluable knowledge on the best time to reclaim fallow land is passed down through generations via poems and songs.

Fig. 7.7 The *p'dav* (*Macaranga denticulata*) growing in fallow forest, Hin Lad Nai, Chiang Rai, Thailand. *P'dav* are trees with large broad leaves (Photo by Prasert Trakansuphakon/PASD)

The *p'dav* is a tree species that has a soft trunk and branching roots that spread at shallow depths around the stump. It only propagates with seeds, which can be collected during August and September. Several varieties are known to local communities: 1) *p'dav Caiv*; 2) *p'dav gauz hpo* (red *p'dav*); 3) *p'dav wa* (white *p'dav*); and 4) *p'dav loz loz* (normal *p'dav*). Although it is easier to collect the seeds of red and white *p'dav*, the normal *p'dav* is better for regenerating soil.

The fallen leaves of *p'dav* trees keep the soil surface cool and hold rainwater, and decay quickly and thoroughly, becoming good, fertile soil. Accordingly, the soil of *p'dav* fallows becomes suitable for farming earlier (in 5–6 years) than that of normal fallows. Farmers know that the soil has improved if it is soft, black, and full of earthworms. The presence of holes made by wildlife, such as bamboo rats (genus *Rhizomys*), porcupines, and wasps, is another indication.

A farmer, Mr. Somboon Siri, compared fallow lands with *p'dav* (3 plots) and without *p'dav* (3 plots) for 14 years (SwedBio and PASD 2016, pp. 9–11). His observations suggest that the fallow lands with *p'dav*:

1. Were more suited for cultivation.
2. Had fewer grasses unfavourable to soil fertility, e.g. *Nauf ne si*, *Nauf k'seij has*, *k'po hpo* (local names).
3. Had healthier plants and larger harvests of almost all crops (e.g. rice production increases)—for example, for 1 kg of rice seed input, 35 kg of rice was harvested after *p'dav* fallow vs. 30 kg after no-*p'dav* fallow.
4. Had more reliable production under varying seasonal weather (dry or wet).

It is believed that shade created by the *p'dav* trees helps to control most of the grasses which make the soil poor for crop cultivation. Some of these grasses are considered weeds for rice and other plants in the rice field.

The *p'dav* produces large quantities of fruit, which attracts wildlife[5] and aids dispersal and regeneration. The *p'dav* areas, therefore, become productive hunting and trapping grounds. Domesticated animals, e.g. buffaloes and cows, favour the *p'dav* fruit, making *p'dav* forests foraging areas for them as well.

Because the *p'dav* grows fast, and its big leaves shade the ground underneath, there is a possibility that the preferential use of the *p'dav* could decrease biodiversity in the fallow lands. Empirical studies are required to conclude the debate over the benefits and drawbacks of the use of *p'dav*.

In parallel, there is a discussion on whether it might be effective to use the *p'dav* for recovering impoverished soil after intensive cash crop cultivations. For instance, farmers could plant shade crops (e.g. coffee or tea) under *p'dav*. Also, in areas where people need firewood to cook, the *p'dav* could be helpful because its branches make good firewood, and the tree grows fast. Trees are good for firewood in just 2–3 years, and they are easy to dry, light to carry, and easy to cut into pieces to burn.

Other benefits of rotational farming include production of plant species for local consumption and potential income, e.g. *Litsea cubeba* (the lemongrass tree or *s'luj saf* in Karen) and *Rhus chinensis* (the Chinese sumac). Beekeeping in fallows contributes to both livelihoods (honey production and pollinating crops) and biodiversity conservation. Tree cover in fallows provides habitats for wildlife species. Older fallow forests (7 years and older) have abundant vines, which are used for herbal medicine and handicrafts. Thus, fallow land is hardly "wasteland".

7.4 Discussion

The area that is cultivated in any given year is a "unit", and there are several of these units under a TRA system; the number of units corresponds approximately to the number of years of the TRA cycle. Ecosystems within individual units are cleared and burned, which is the source of the commonly held conception that TRA is harmful to biodiversity. However, TRA should be considered in a more macro

---

[5] Examples include:

- Birds: *htof bau lauv* (red-whiskered bulbul, *Pycnonotus jocosus*), *htof bau hpo, bif bei* (minivets, *Pericrocotus* spp.), wild chicken, *htof bgef* (Siamese partridge, *Arborophila diversa*), *htof giv* (pheasants), and at least five different species of dove (*htof lwij bu* [rice dove], *htof lwij gauz* [red dove], *htof lwij htau maij* [long-tail dove], *htof lwij qei* [dry dove], *htof lwij hkaf* [bitter dove]); direct translation of local names).
- Rats and squirrels: *Liv lai, yuj lai, liv bau hkaf, liv looj gauz sav, yuj qoov yuj gauz, yuj hko*.
- *Taj hpo nyau* (the palm civet).
- Bees: *kwaiv dof* (eastern honey bee, *Apis cerana*), *kwaiv hpo* (meliponines), *de t'yoo* (dwarf honey bee, *Apis florea*) etc.

view—at the level of a landscape rather than in terms of each "unit". At the landscape level, a range of different but interacting ecosystems and land uses are supported, including mixed crop fields, pasturelands, regenerating forests of varied ages, and primary or old successional forests. This landscape-level system is buttressed with other land uses, particularly home gardens and rice cultivation in the valleys. Furthermore, human use, or anthropogenic disturbance, is contained within the TRA system, and surrounding forest is kept intact whilst farming is rotated across fixed patches—where and how farming has been practiced traditionally.

As the economic and demographic circumstances, as well as the political discourses on TRA, change, the practice of TRA needs to adapt to maintain benefits and minimise shortcomings (Table 7.1). Below, we discuss the future of TRA from environmental and socio-economic perspectives, as they relate to ecosystem restoration.

7.4.1   Environmental Perspectives

A long fallow period as traditionally practiced has the environmental benefits of: (1) storing more carbon, an important function forests play in mitigating climate change; (2) providing diverse habitat types at different stages of ecological succession (Fig. 7.3) with higher biodiversity than both short fallows (e.g. Wang and Young 2003) or alternative land uses such as palm oil and teak plantations (e.g. Mandal and Raman 2016); (3) harbouring high levels of agro-biodiversity through cultivation of locally adapted, traditional crop varieties (e.g. Rerkasem et al. 2009; Amba Jamir, pers. comm.); (4) controlling weed growth and regenerating soil fertility from biomass accumulation and nutrient cycling (e.g. Teegalapalli et al. 2018); and (5) regulating water flows and soil erosion (e.g. Ziegler et al. 2009), amongst others. Moreover, paleoecological research suggests that shifting cultivation systems of varied fallow lengths enhance both landscape-scale biodiversity and the resilience of forests to climate shifts (Hamilton et al. 2020), and have played a role in shaping some of the most biodiverse regions of the planet (Fletcher et al. 2021).

The two cases presented above demonstrate that the use of traditional practices allows for many environmental benefits, even with short fallow cycles. The contribution of TRA to the economic and livelihood security of upland communities is undeniable, as are contributions to forest cover maintenance. Effective and productive uses of short fallow periods based on the economic, ecological, and cultural requirements of the local people, ranging from agroforestry plantations to the alder, *p'dav*, and similar management practices, are beneficial, especially considering the close cultural affinities with traditional practices.

Although the *p'dav* and the alder are featured here, there are many other species that can be used to regenerate fertility during the fallow phase. This model of ecosystem restoration can be replicated elsewhere by identifying and using species suited to the area of interest. Moreover, the use of intelligent design principles—e.g.

practices like leaving hilltop forests untouched as water catchments; situating culti-vated fields close to primary or late-successional forests to facilitate nutrient and water flows, seed dispersal, and pollination (Teegalapalli et al. 2009); interspersing community-conserved areas; and regulating hunting of wildlife that aid in forest recovery—will ensure both ecosystem restoration and food security.

Raising pollinators for honey and pollination services can address conservation challenges in TRA in multiple ways. As briefly mentioned in the discussion of Hin Lad Nai, beekeeping in fallow lands is an innovation that directly and indirectly supports biodiversity. Bees pollinate plants in the fallow land, and in forests and grasslands around the fallow lands, benefiting the entire ecosystem. Besides the important function of pollination, bees also produce honey for consumption and income for the farmers. Since bee boxes are kept in the fallow lands, beekeeping also motivates farmers to take care of their fallow lands and forest areas. By keeping bees, the local farmers take care of the whole ecosystem, simultaneously securing their food supply and income from non-timber forest products (NTFPs).

7.4.2 Socio-Economic Perspectives

Despite pressures to eliminate shifting cultivation, the practice continues to be well-entrenched in both Northeast India (e.g. Pandey et al. 2020) and in Northern Thailand. Almost 90% of a sample of Adi families interviewed in central Arunachal Pradesh, India, continue to practice shifting cultivation despite the introduction of settled cultivation in the 1960s Teegalapalli and Datta 2016). A recent study of 52 northeastern villages in India (Pandey et al. 2020) found that shifting cultivation continued to be communities' preferred mode of cultivation primarily because of shared communal ties and traditions (social bonding), and their dependence on *jhum* as a means of livelihood rooted in their remote mountainous homes in the absence of other viable alternatives (Sarma 2022; Bhattacharjee 2022) for food production.

Jhum systems interwoven with the elaborate social and cultural traditions of tribal communities provide food safety nets for poorer households. Amongst the Sümi, Konyak, and Chang tribes of Nagaland, the chief who owns most of the land decides who gets a piece of land to farm based on the needs of each member of the village community. This and other highly divergent tribal land ownership systems ensure that all members of the village community have land to cultivate and people are not left "landless". Shifts to more permanent cropping systems owned and managed by individuals have negative impacts on the socio-economic security of *jhum* cultiva-tors, and on community decision-making for sustainable land management (e.g. Das 2020). These shifts might also impact the range of products and services that the landscapes provide (such as water, NTFPs, wildlife to hunt, timber, fuel, and crop diversity), decreasing the resilience of these socio-ecological production landscapes and their regenerative capacity.

Transitions to settled cultivation are also likely to undermine the role that women play in protecting agro-biodiversity, and eventually the influence they wield in ensuring crop resilience. Women were traditionally in charge of seed management and exchange, carrying seeds between their natal and marital villages (Goodrich 2012). Altering TRA practices could therefore have unforeseen effects in multiple dimensions.

7.4.3   Opportunities and Challenges

TRA can be improved and adapted to changing circumstances. Because TRA uses the regenerative function of natural systems, aspects of ecosystem restoration are built into the traditional practices. Since both *Macaranga denticulata* and *Alnus nepalensis* are common species in the region, opportunities to replicate the practices of communities in Hin Lad Nai and Khonoma villages would be possible with some effective outreach and extension activities. A programme to support farmer-to-farmer information exchange, such as exposure visits, could facilitate such capacity building.

TRA can also employ established innovations in ecosystem restoration during both cultivation and fallow periods to better cope with environmental and economic challenges. Examples include soil conservation measures, reforestation of fallow plots combined with agroforestry, and assisted forest regeneration with strategically selected tree species. As highlighted in the Hin Lad Nai case, NTFPs such as honey can contribute positively not only to the livelihoods of the farmers, but also to the care of forests in fallow periods. A win-win relationship can be created between environmental and economic concerns by focusing on the productive management of fallow forests. Despite the foreseeable benefits, initial investments to put new activities into motion, as well as motivating farmers to take on new activities, continue to pose challenges.

7.5   Conclusion

In conclusion, TRA is dynamic and many solutions exist for the rapid regeneration of short fallows, but all must be based on traditional knowledge, practices, and philosophy. TRA, consisting of cycles of cultivation and fallow periods, is an expression of indigenous self-sufficiency and sustainable natural resource management. The cultural, social, and religious fabric of tribal communities in Northeast India and Northern Thailand is intricately connected to this practice of land use. Most of the traditional festivals correspond to various stages of cultivation. The rejection of TRA in favour of settled cultivation and/or plantations could therefore undermine traditional social cohesion with possible negative impacts on the social-ecological resilience of these forest-farm landscapes. Ancient wise-use taboos, along with indigenous and local knowledge and practices (ILKPs), contributed to environmental sustainability in the past, but are eroding today. The diminution of social cohesion and traditional knowledge held amongst the community members leads to the loss of knowledge that can inform ecosystem restoration. Reviving where appropriate, and learning from these ILKPs (e.g. alder management of the Angamis or *p'dav* of the Karen people), contributes to ensuring (sensu Dasgupta et al. 2021) that sustainable environmental boundary conditions persist.

Acknowledgements   The documentation of activities in Khonoma village in India was supported by a grant from the Japan Fund for Global Environment. We thank various people, particularly Mr. Amba Jamir, Dr. Anirban Datta-Roy, and Mr. Angulie Meyase, for sharing their knowledge and providing several important references for this chapter.

References

Agrawal A, Kumar V (2012) An investigation into changes in Nagaland's population between 1971 and 2011. In: IEG Working paper No. 312. Institute of Economic Growth, University of Delhi, New Delhi

Bruun TB, de Neergaard A, Lawrence D, Ziegler AD (2009) Environmental consequences of the demise in Swidden cultivation in Southeast Asia: carbon storage and soil quality. Hum Ecol 37: 375–388

Bhattacharjee S (2022) Re-assessing Jhum in Arunachal Pradesh: colonial perceptions, indigenous environmental practices and post-colonial transitions. In: Shaw AK, Natori Y, Edake S (eds) A tradition in transition: understanding the role of shifting cultivation for sustainable development of north East India. TERI Press, New Delhi, pp 106–125

Borah D, Goswami NR (1973) A comparative study of crop production under shifting and terrace cultivation: a case study in Garo hills, Meghalaya. Ad hoc Study No. 35. Agro-economic Research Centre for North East India, Assam Agricultural University, India

Cairns M (2007) The alder managers: the cultural ecology of a village in Nagaland, N.E. India. Ph. D. thesis, Australian National University, Canberra

Cairns M, Brookfield H (2011) Composite farming systems in an era of change: Nagaland, Northeast India. Asia Pac Viewp 52:56–84. https://doi.org/10.1111/j.1467-8373.2010.01435.x

Cairns M, Keitzar S, Yaden TA (2007) Shifting forests in Northeast India. Management of *Alnus nepalensis* as an improved fallow in Nagaland. In: Cairns M (ed) Voices from the Forest: integrating indigenous knowledge into sustainable upland farming. Resources for the Future, New York, pp 341–378

Das D (2020) Modernity lacks care: community-based development and the moral economy of households in eastern Nagaland. J South Asian Dev 15:97–116

Dasgupta R, Dhyani S, Basu M, Kadaverugu R, Hashimoto S, Kumar P, Johnson BA, Takahashi Y, Mitra BK, Avtar R, Mitra P (2021) Exploring indigenous and local knowledge and practices (ILKPs) in traditional jhum cultivation for localizing sustainable development goals (SDGs): a case study from Zunheboto district of Nagaland, India. Environ Manag:1. https://doi.org/10. 1007/s00267-021 01514-6

Finegan B, Nasi R (2004) The biodiversity and conservation potential of shifting cultivation landscapes in tropical landscapes. In: Schroth G, da Fonseca GAB, Harvey CA, Gascon C, Lasconcelos HL, Izac A-MN (eds) Agroforestry and biodiversity conservation in tropical landscapes. Island Press, Washington, DC, pp 153–197

Fletcher MS, Hamilton R, Dressler W, Palmer L (2021) Indigenous knowledge and the shackles of wilderness. Proc Natl Acad Sci U S A 118(40):1–7

Gogoi A, Sahoo UK, Saikia H (2020) Vegetation and ecosystem carbon recovery following shifting cultivation in Mizoram-Manipur-Kachin rainforest eco-region, Southern Asia. Ecol Proc 9:1. https://doi.org/10.1186/s13717-020-00225-w

Gokhale AM, Zeliang DK, Kevichusa R, Angami T, Bendangnungsang S (1985) The use of alder trees. State Council of Educational Research and Training, Education Department, Government of Nagaland, Kohima

Goodrich GC (2012) Gender dynamics in agro-biodiversity conservation in Sikkim and Nagaland. In: Krishna S (ed) Agriculture and a changing environment: perspectives on north-eastern India. Routledge, New Delhi

Hamilton R, Penny D, Hall TL (2020) Forest, fire & monsoon: investigating the long-term threshold dynamics of south-East Asia's seasonally dry tropical forests. Quatenary Science Reviews 238: 106334. https://doi.org/10.1016/j.quascirev.2020.106334

Hutton JH (1921) The Sema Nagas. MacMillan and Co. Ltd., London

Kehie M, Khamu S, Kehie P (2017) Indigenous alder based farming practices in Nagaland, India. J Traditional Folk Pract 5(2):82–152

Lamichhaney BP (1995) *Alnus nepalensis* D. Don: A detailed study. In *FORESC Monograph 1*, Forest Research and Survey Centre, Ministry of Forests and Soil Conservation, Kathmandu

Mandal J, Raman TRS (2016) Shifting agriculture supports more tropical forest birds than oil palm or teak plantations in Mizoram, Northeast India. Condor 118:345–359. https://doi.org/10.1650/CONDOR-15-163.1

Maithani BP (2005) Shifting cultivation in north-East India: policy issues and options, Mittal Publications, New Delhi, viewed 10 February 2022. Retrieved from https://books.google.co.jp/books?printsec=frontcover&vid=LCCN2005322020&redir_esc=y#v=onepage&q&f=false

Meena V (2018) GIS file of India State, *District and Tehsil Boundaries*, map, ArcGIS Hub, accessed on 31 August 2021. Retrieved from https://hub.arcgis.com/content/cba8bddfa0ab43ddb35a7313376f9438/

NEPED & IIRR (1999) Building upon traditional agriculture in Nagaland, India. Nagaland Environment Protection and Economic Development & International Institute of Rural Reconstruction, Silang

NRSC & DOLR (2005–2019) Wasteland atlas of India, National Remote Sensing Centre, Indian Space Research Organisation, Department of Space Hyderabad and Department of Land Resources, Ministry of Rural Development, Government of India, New Delhi, viewed 4 February 2022. Retrieved from https://dolr.gov.in/documents/wasteland-atlas-of-india

Pandey DK, De HK, Dubey SK, Kumar B, Dobhal S, Adhiguru P (2020) Indigenous people's attachment to shifting cultivation in the eastern Himalayas, India: a cross-sectional evidence. Forest Policy Econ 111:102046. https://doi.org/10.1016/j.forpol.2019.102046

Pant PM, Panchayati R, Tiwari BK, Choudhury D (2018) Shifting cultivation: towards a transformational approach. Report of Working Group III, NITI Aayog, New Delhi, viewed 4 February 2022. Retrieved from https://lib.icimod.org/record/34338

Ramakrishnan PS (1992) Shifting agriculture and sustainable development: an interdisciplinary study from North-Eastern India. *MAB Series*, vol. 10. UNESCO, Paris

Ranjan R, Upadhyay VP (1999) Ecological problems due to shifting cultivation. Curr Sci 77:1246

Rathore SS, Karunakaran K, Prakash B (2010) Alder based farming system a traditional farming practices in Nagaland for amelioration of *jhum* land. Indian J Tradit Knowl 4:677–680

Rerkasem K, Lawrence D, Padoch C, Schmidt-Vogt D, Ziegler AD, Bruun TB (2009) Consequences of swidden transitions for crop and fallow biodiversity in Southeast Asia. Hum Ecol 37: 347–360

Royal Thai Survey Department (2019) Thailand (THA) Administrative Boundary Common Operational Database, map, Humanitarian Data Exchange, viewed 19 February 2022. Retrieved from https://data.humdata.org/dataset/cod-ab-tha

Sarma A (2022) Can shifting cultivation ensure local food sovereignty? In: Edake S (ed) A tradition in transition: understanding the role of shifting cultivation for sustainable development of north East India. TERI Press, New Delhi

Srinivasan U (2014) Oil palm expansion: ecological threat to north-East India. Econ Political Weekly, 49(36), viewed 10 February 2022. Retrieved from http://www.epw.in/journal/2014/3 6/reports-states-web-exclusives/oil-palm-expansion.html

SwedBio & PASD (2016) Mobilizing Traditional Knowledge, Innovations and Practices in rotational farming for sustainable development, SwedBio at Stockholm Resilience Centre and Pgaz K' Nyau Association for Sustainable Development, viewed 10 February 2022. Retrieved from https://swed.bio/wp-content/uploads/2016/11/MEB-Pilot-Report-Thailand_2016.pdf

Teegalapalli K, Datta A (2016) Shifting to settled cultivation: changing practices among the Adis in Central Arunachal Pradesh, north-East India. Ambio 45:602–612. https://doi.org/10.1007/s13280-016-0765-x

Teegalapalli K, Veeraswami GG, Samal PK (2009) Forest recovery following shifting cultivation: an overview of existing research. Trop Conserv Sci 2:374–387

Teegalapalli K, Mailappa AS, Lyngdoh N, Lawrence D (2018) Recovery of soil macronutrients following shifting cultivation and ethnopedology of the Adi community in the eastern Himalaya. Soil Use Manag 34:249–257

Wang Z, Young SS (2003) Differences in bird diversity between two swidden agricultural sites in mountainous terrain, Xishuangbanna, Yunnan, China. Biol Conserv 110:231–243. https://doi.org/10.1016/S0006-3207(02)00222-7

Ziegler AD, Bruun TB, Guardiola-Claramonte M, Giambelluca TW, Lawrence D, Lam NT (2009) Environmental consequences of the demise in Swidden cultivation in montane mainland Southeast Asia: hydrology and geomorphology. Hum Ecol 37:361–373. https://doi.org/10.1007/s10745-009-9258-x

The opinions expressed in this chapter are those of the author(s) and do not necessarily reflect the views of UNU-IAS, its Board of Directors, or the countries they represent.

Open Access This chapter is licenced under the terms of the Creative Commons Attribution-NonCommercial-ShareAlike 3.0 IGO licence (http://creativecommons.org/licenses/by-nc-sa/3.0/igo/), which permits any noncommercial use, sharing, adaptation, distribution and reproduction in any medium or format, as long as you give appropriate credit to UNU-IAS, provide a link to the Creative Commons licence and indicate if changes were made. If you remix, transform, or build upon this book or a part thereof, you must distribute your contributions under the same licence as the original. The use of the UNU-IAS name and logo, shall be subject to a separate written licence agreement between UNU-IAS and the user and is not authorised as part of this CC BY-NC-SA 3.0 IGO licence. Note that the link provided above includes additional terms and conditions of the licence.

The images or other third party material in this chapter are included in the chapter's Creative Commons licence, unless indicated otherwise in a credit line to the material. If material is not included in the chapter's Creative Commons licence and your intended use is not permitted by statutory regulation or exceeds the permitted use, you will need to obtain permission directly from the copyright holder.

Chapter 8
Restoring Rice Paddies and Rice Agro-Ecosystem Services Through a Participatory Seed Conservation and Exchange Programme

Archana Bhatt, N. Anil Kumar, C. S. Dhanya, and P. Vipindas

8.1 Introduction

Rice has been a staple crop in the majority of Asian nations since time immemorial and has evolved itself within various landscapes and food patterns that span borders. Rice agro-ecosystems across Asia are highly diverse in terms of varieties, cultivation practices, and food patterns, depending on the regions and community spread throughout the entire continent. This diversity also allows high nutritional benefits and helps ensure the food and nutritional security of communities (Zeng et al. 2010; Umadevi et al. 2012).

In the era of the Green Revolution of the 1960s, revolutionary steps were made towards combating hunger through the introduction of High Yielding Varieties (HYVs) of rice and wheat, coupled with the use of fertilisers and machinery. But along with the benefits came the decline in traditional paddy varieties (Parayil 1992; Eliazer Nelson et al. 2019; Roy et al. 2018). India alone was once home to around 200,000 landraces of rice (Richharia and Govindasamy 1990), many of which formed part of the cuisine and various religious rituals as reported in the ancient texts of Ayurveda (Jose et al. 2018; Sathya 2013). The HYVs, combined with increased and uneven use of fertilisers and pesticides, added up to the loss of many other organisms in the food web of the rice agro-ecosystem (Kumar 2017). Moreover, extreme weather events, especially delayed monsoons, droughts, and floods due to untimely heavy rains have become a major factor contributing to the decline in rice cultivation over the past decade. Various studies reflect that the productivity of many crops, including rice, is expected to be significantly reduced

A. Bhatt (✉) · N. A. Kumar · C. S. Dhanya · P. Vipindas
MSSRF-Community Agro biodiversity Centre, Puthoorvayal, Meppadi, Kerala, India
e-mail: archanab@mssrf.res.in

© The Author(s) 2023
M. Nishi, S. M. Subramanian (eds.), *Ecosystem Restoration through Managing Socio-Ecological Production Landscapes and Seascapes (SEPLS)*, Satoyama Initiative Thematic Review, https://doi.org/10.1007/978-981-99-1292-6_8

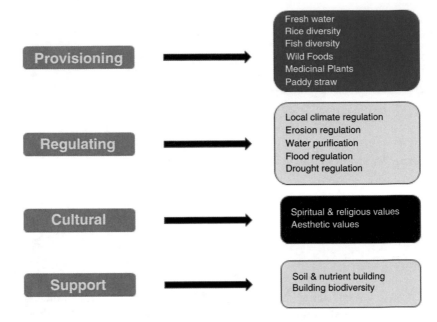

Fig. 8.1 Ecosystem services of the rice paddy agro-ecosystem (Source: Prepared by authors based on Millennium Ecosystem Assessment (2005))

in the coming decades due to increased climate variability and climate change (Salinger et al. 2005; Godfray et al. 2010; van Noordwijk et al. 2018).

Paddy agro-ecosystems generate a number of services (Chivenge et al. 2020) and support diverse species that use the paddy fields as their habitat or food source. In terms of biodiversity and ecosystem services, a wetland paddy field provides various aspects of provisioning, regulating, supporting, and cultural services for communities (Fig. 8.1). A typical rice agro-ecosystem is home to many other species apart from the rice that is cultivated in the field. A study conducted by Parameswaran and Kumar (2017) reported 95 plant species in a paddy agro-ecosystem (dominated by the members of Cyperaceae, Asteraceae, Poaceae, Amaranthaceae families) that were of direct use value to communities as wild edible greens or medicinal herbs. Rice paddies also provide habitats for birds, insects, fish, and other forms of diversity, which in turn helps to reduce incidences of pest damage (Kumar et al. 2022). A small field of paddy rice is actually a complex web of diversity woven into an agro-ecosystem that supports various life forms. Therefore, a decline in rice cultivation and the diversity of varieties cultivated eventually results in a threat to ecosystem services in their entirety.

Furthermore, traditional crop production practices help in maintaining agro-biodiversity, which in turn also helps in conserving the associated cultural wisdom important for crop productivity (Cox 2000; Maffi 2002).

In 2009, the Convention on Biological Diversity (CBD) defined "Ecosystem-based Adaptation (EbA)" as "the use of biodiversity and ecosystem services as part of an overall strategy to help people adapt to the adverse effects of climate change" (CBD 2009, p. 31). There are three core elements of EbA: use of biodiversity and ecosystem services, the dependent communities, and adaptation to climate change. Considering the eco-sociological importance of the rice production system, a Rice Ecosystem-based Adaptation (REbA) approach becomes important for conserving variability in the rice crop and the associated biodiversity, as well as capacity development of rice farming communities for adaptation to climate change.

Since the livelihoods of rural and indigenous communities are threatened by climate change (Easterling et al. 2007), it is of primary importance to strengthen the adaptation capacity and resilience of such communities. Wayanad, one of the hilly districts of the state of Kerala in India, has the highest population of indigenous or tribal communities in the state. These communities are engaged in cultivation of traditional paddy varieties and conservation of crop diversity.

8.2 The Intervention Site: Wayalnadu—"Land of Rice Fields"

Kerala, popularly known as "God's Own Country" on account of its scenic landscapes, lies in the southernmost part of the Indian subcontinent and is home to the Western Ghats, one of the major biodiversity hot spots of the world. The state has a diverse topography ranging from vast wetlands, the picturesque coastline, to the foothills of the Western Ghats that provide a habitat for the varied germplasm of wild and domesticated flora. Cultivation of rice in the state dates back to 5000 BC (Manilal 1991), and rice holds an important place as the staple food of people. But this diversity, along with the area of paddy crops under cultivation, has been declining over the years. The area of paddy cultivation in Kerala has dropped from 2,100,396 acres in 1980–1981 to a mere 489,713 acres in 2019–2020 (GoK 2019) due to a range of factors. Major ones include alternative land uses, shifts to other input-intensive crops, lower productivity, disease and pest incidence, labour shortages, and climate variability (Hari and Kumar 2016; Fox et al. 2017). This decline in rice cultivation has not only impacted grain production, but has also adversely affected the ecosystem services offered by these unique ecosystems (Kumar and Kunhamu 2021).

Physiographically, Wayanad comes under the category of eastern highlands (a cool hilly terrain) also known as Malanadu. Wayanad is a microcentre of rice crop diversity, part of the Western Ghats' global biodiversity hotspot, and home to the majority of indigenous communities in Kerala. The name Wayanad literally translates to Wayalnadu, meaning "land of rice fields", and the rice agro-ecosystem in the region consists of typical homestead-based farmlands divided into three zones of cultivation. This zoning is primarily based on the topographic conditions of the land, i.e. lowland paddy fields, a homestead and mixed tree-crop system in the vicinity, and tree crops in the upland area (plantation and spice crops). All zones

Table 8.1 Basic information of the study area

Country	India
Province	Kerala
District	Wayanad
Size of geographical area (hectare)	213,100
Number of direct beneficiaries (persons)	51,320
Number of indirect beneficiaries (persons)	817,420
Dominant ethnicity(ies), if appropriate	Hindu
Size of the case study/project area (hectare)	134,313
Geographic coordinates (latitude, longitude)	11.6854° N, 76.1320° E

(Source: Census of India District Handbook 2011)

contribute a range of ecosystem services, ranging from food, grazing for livestock, soil conservation, maintenance of groundwater table, and habitats for associated floral and faunal biodiversity. This system strongly resembles the Satoyama landscapes of Japan, which evolved over centuries through long-term interaction between human beings and their local environments (Kumar and Takeuchi 2009).

A study helmed by the M S Swaminathan Research Foundation (MSSRF) in 2011 (GoK 2011) showed the existence of more than 75 traditional rice varieties cultivated in the Wayanad district, whilst another study conducted in the region reported only 60 traditional paddy varieties (KAU 2018). Most of these traditional rice varieties are cultivated by tribal farmers, but cultivation is declining due to a range of factors. Various studies and reports have emphasised that when rural communities are engaged in the management of ecosystem services, the sustainability of the ecosystems can be enhanced alongside mitigation of the effects of climate change and improvement in livelihoods (FAO 2009). Keeping in view the role of the rural and indigenous community, MSSRF proposed a holistic approach to support ecosystem services and conservation of biodiversity in the paddy agro-ecosystem in Wayanad (Table 8.1). Discussions in this regard led to identifying various challenges associated with traditional paddy cultivation in the region (Fig. 8.2) and building awareness on an EbA approach with the traditional paddy agro-ecosystem and Rice Seed Village (RSV) as the critical components.

8.3 Methodology

A participatory rice seed production programme was initiated in 2011 with the objectives of conservation of the maximum possible rice crop genetic diversity and the building of capacity amongst rice cultivators to adapt to climate change and related vagaries in five villages of Wayanad district. The number of sites has increased to 30 villages as of 2021 (Fig. 8.3). The intervention started with documentation of traditional knowledge and practices, focusing on mobilisation of

Fig. 8.2 Challenges associated with traditional paddy cultivation in the region (Source: Prepared by authors based on data collected from seed villages)

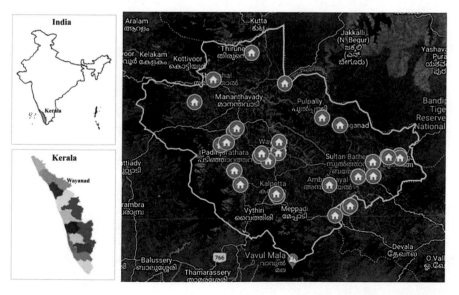

Fig. 8.3 Map indicating area of intervention and seed villages (Source: Prepared by authors with ArcGIS and Google Earth)

farming families that were cultivating rice and were interested in cultivation of Traditional Rice Varieties (TRVs). The intervention followed a step-by-step approach starting with community mobilisation, development of a farmers' inventory, and problem identification, and moving on to training support in Good Agricultural Practices (GAP), quality seed production techniques ranging from seed selection to purification and storage, and facilitating market linkages. Necessary linkages were developed amongst the established seed villages and fellow farmers, marketing channels, agricultural officers, consumers, and other stakeholders. The results of the interventions were synthesised through various qualitative data collection methods, mainly semi-structured interviews, field observation, focus-group discussions, and key informant interviews. Resulting data was then analysed and interpreted.

Over a 10-year period, a range of interrelated activities covering the three thematic areas of biodiversity conservation, climate change adaptation, and livelihood development were undertaken (Fig. 8.4). The following three major activities were streamlined for concerted action.

8.3.1   Farmers' Training on Climate-Resilient Agricultural Practices

Starting in 2015, 27 training programmes have been conducted covering the selected subject areas, reaching 459 male and 274 female farmers from the RSVs. Trainings were provided on various climate-resilient practices—predominantly System of Rice Intensification (SRI) and Modified SRI (MSRI)—that helped the farmers to gain necessary skills in adapting to drought conditions. SRI is a climate-resilient practice of paddy cultivation developed in Madagascar and followed worldwide on account of the benefits it offers with respect to productivity and climate change (Thakur and Uphoff 2017). The adoption of SRI and MSRI by the RSV farmers allowed them to cultivate paddy under water scarcity and also helped them gain higher yields (20%) compared to the conventional method of rice cultivation. Training on preparation and usage of various organic nutrient formulations, like *Jeevamrutham, Beejamrutham, Panchgavyam*, and Fish Amino, also aided yield enhancement without harmful chemicals. Various other climate change adaptation practices when promoted amongst the community helped the farmers to mitigate and adapt to the effects of climate variability. These mainly included: cultivation of pulse or legume crops in the rice fields after the harvest; green manuring; application of soil nutrients in split doses; soil test-based nutrient application; change in sowing or transplanting dates in case of delayed monsoons; and establishment of nurseries in uplands. Floods in a way also revived the knowledge and value of traditional varieties amongst the new generation of farmers.

As part of the programme, indigenous knowledge documentation revealed various insights on the management practices followed in the past by paddy farmers, which are detailed below. This documented knowledge was validated during the

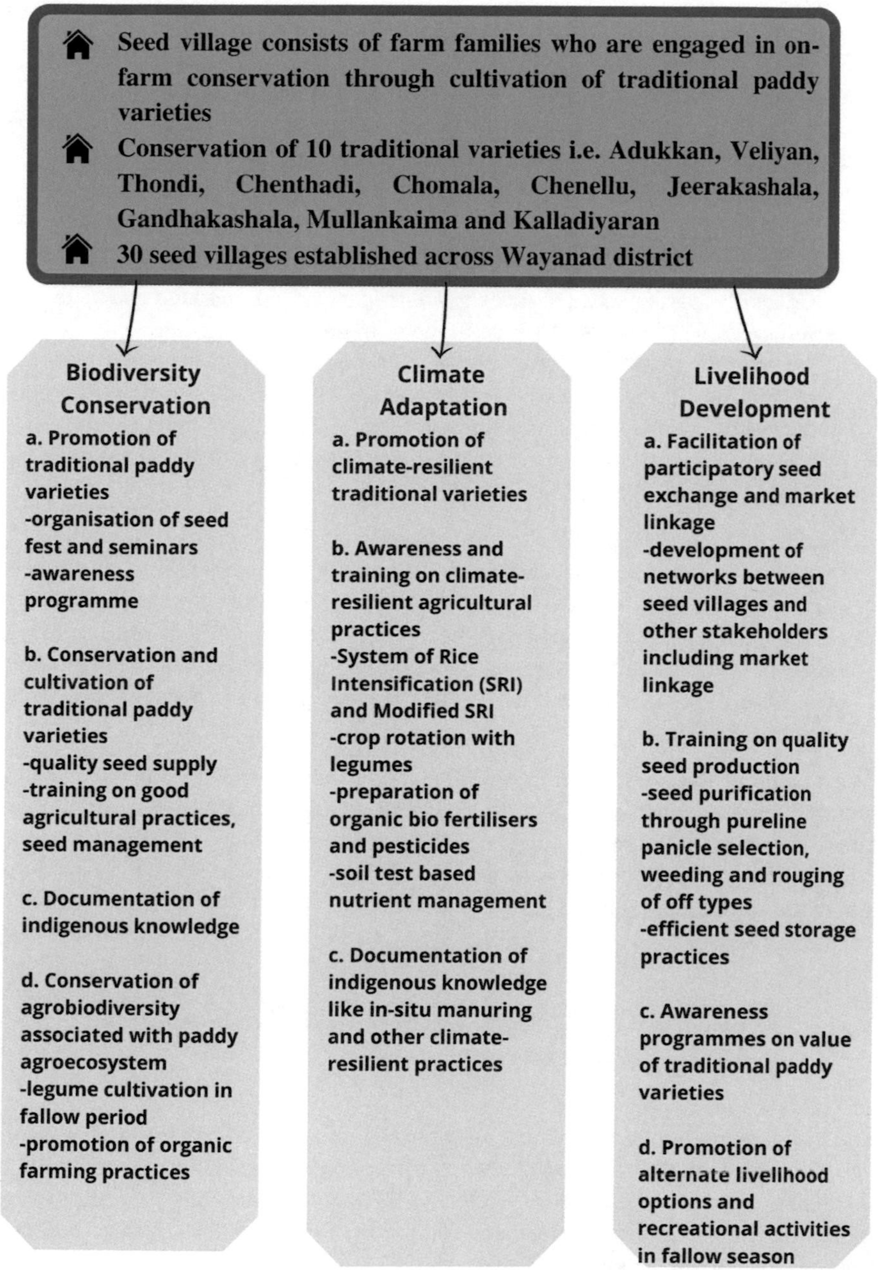

> 🏠 Seed village consists of farm families who are engaged in on-farm conservation through cultivation of traditional paddy varieties
>
> 🏠 Conservation of 10 traditional varieties i.e. Adukkan, Veliyan, Thondi, Chenthadi, Chomala, Chenellu, Jeerakashala, Gandhakashala, Mullankaima and Kalladiyaran
>
> 🏠 30 seed villages established across Wayanad district

Biodiversity Conservation

a. Promotion of traditional paddy varieties
-organisation of seed fest and seminars
-awareness programme

b. Conservation and cultivation of traditional paddy varieties
-quality seed supply
-training on good agricultural practices, seed management

c. Documentation of indigenous knowledge

d. Conservation of agrobiodiversity associated with paddy agroecosystem
-legume cultivation in fallow period
-promotion of organic farming practices

Climate Adaptation

a. Promotion of climate-resilient traditional varieties

b. Awareness and training on climate-resilient agricultural practices
-System of Rice Intensification (SRI) and Modified SRI
-crop rotation with legumes
-preparation of organic bio fertilisers and pesticides
-soil test based nutrient management

c. Documentation of indigenous knowledge like in-situ manuring and other climate-resilient practices

Livelihood Development

a. Facilitation of participatory seed exchange and market linkage
-development of networks between seed villages and other stakeholders including market linkage

b. Training on quality seed production
-seed purification through pureline panicle selection, weeding and rouging of off types
-efficient seed storage practices

c. Awareness programmes on value of traditional paddy varieties

d. Promotion of alternate livelihood options and recreational activities in fallow season

Fig. 8.4 Methodology adopted in Rice Seed Villages (Source: Prepared by authors based on pilot study)

training sessions for farmers, who were encouraged to revitalise these practices wherever possible in the fields.

- Communities determine the timing of sowing/transplanting and harvesting of the crop based on *Paurnami*, i.e. the lunar cycle. Further discussion with the farmers and agriculture experts revealed that during certain periods in the lunar cycle, rodents and pests stay away from the field and do not affect the crops.
- Farmers follow a range of eco-friendly management practices to manage diseases and pests in the field, such as:

  - Usage of a particular plant species (*Ficus hispida*) with a rough and spiny leaf surface to ward off insect larvae in the crop field.
  - Attracting owls—the natural predator of pests like rodents—and usage of stuffed owl feathers to avoid rodent attacks in the field, also help to reduce the use of chemicals and maintain the fabric of the natural food web.
  - Usage of cow dung slurry to provide extra nutrients to the crop, reducing dependence on chemical fertilisers.

- Grazing of cattle in the field after harvest was also facilitated with the cultivation of traditional varieties. Since the traditional varieties have more height and thicker straw, they are quite suited to livestock. Hence, the stubble left after the harvest can be easily utilised through grazing, which also promotes in situ manuring. Based on their traditional wisdom, the communities allow equitable distribution of manure by developing borders in the field for grazing and shifting the cattle over time to other fields.

Throughout the programme period, the associated knowledge was documented noting that such valuable wisdom if lost cannot be regained, making it very important to conserve for future farming generations. Researchers such as Deb (2014) have reported that the disappearance of traditional paddy varieties leads to loss of the associated traditional knowledge of farmers as well.

8.3.2   Facilitation of Participatory Seed Purification, Exchange, and Market Linkage

The focus of this activity was to extend the programme's aims past improving yields of rice to provide a strategic framework for conserving, consuming, and gainful marketing of grains and seeds of TRVs. In this context, the following steps were initiated: identifying maximum genetic variability of the varieties known to farmers; joint testing and selection by farmers and the MSSRF team on-site in differing targeted farm environments; and participatory seed exchanges. The TRVs selected for cultivation were determined based on the preferences of the farming communities. Factors included taste, size and shape of the grain, nutritional and aromatic properties, and the drought and/or flood resistance/tolerance of the varieties.

8.3.3   Organising Seed Festivals

As part of the intervention, seed festivals (*Vittutsavam*) were organised annually starting in 2014. These events provided the right platform to share knowledge related to farm agro-biodiversity and to stress its importance. With the participation of communities and support of local self-government and grassroots organisations, the programme remains a one-of-a-kind in the state. Seed festivals have allowed for the exchange of seeds of traditional rice varieties between farmers across the state of Kerala. Seeds of most traditional varieties are not easily available to farmers, but through this programme, farmers from various regions could access the seeds that they had lost over time due to lack of a market for their sale. Each seed festival witnessed the participation of more than a thousand visitors of different backgrounds and expertise. The programme also became a medium for knowledge exchange between the communities and academic experts through seminars and panel discussions. Over the years, knowledge exchange on agro-biodiversity governance, the need for community seed banks, resilient systems, networking, farmers' rights, and other policy issues has been facilitated through the event, forming a farmer–scientist–official interface. Farmers engaging in conservation across the RSVs have showcased their farm diversity through exhibitions. Likewise, an interface between farming communities and other stakeholders like traders, the general public, students, and government officials have helped generate awareness on the rich genetic diversity of the region.

8.4   Results

Through the RSV programme, efforts were made to empower the communities through various means, such as participatory seed conservation and exchange, facilitating market linkage, introduction of alternate livelihood options, food and livelihood security, and awareness and knowledge creation. With REbA, the programme directly reached out to 340 farm families in the RSVs and more than 3000 farmers through various activities. The conservation of rice genetic diversity through these villages has aided the socio-economic development of over 300 farming families engaged in the cultivation of traditional rice varieties.

The RSV intervention helped to develop the capacity of farmers in maintaining the rice farming system ecology and to better use surplus seeds and grains with the new partnership established between other farmer groups and the larger rice markets. This is evidenced at present in the functioning of 30 RSVs in the Wayanad district (Table 8.2). The major outcomes of the intervention are described below.

Table 8.2 Details of the RSVs along with names of TRVs and area cultivated

S. No.	Rice Seed Village	Area (in acres) under paddy cultivation	Varieties conserved and cultivated
1	Tengakolli	8	Veliyan, Chenellu, Thondi, Kaima, Jeerakashala, Kunjootti Matta
2	Kolichal	15	Karuvalicha, Mannuveliyan, Gandhakashala, Mullankaima
3	Vaduvanchal	2	Chenthadi, Mullankaima
4	Chethalayam	8	Valichoori, Chenellu
5	Oonimoola	1	Thonooramthondi
6	Chekkady	6	Chenellu, Jeerakashala
7	Alathur	40	Thondi, Chenellu, Gandhakashala, Kunjootti Matta
8	Kolur	2	Thondi, Gandhakasala
9	Cheeyambam	2	Chenellu, Adukkan, Gandhakashala
10	Ambalavayal	16	Kalladiyaran, Oonavattan, Adukkan, Gandhakashala, Veliyan, Mullankaima, Paalthondi, Marathondi, Kottathondi
11	Kallanchira	18	Adukkan, Veliyan, Chenthadi, Chomala, Chenellu, Thondi, Palthondi, Jeerakashala, Gandhakashala, Mullankaima, Kalladiyaran, Njavara
12	Palliyara	10	Veliyan, Thondi, Chenellu, Gandhakashala
13	Kanancheri	1.25	Gandhakashala, Veliyan
14	Ediyamvayal	14	Kalladiyaran
15	Peruvadi	25	Veliyan, Thondi, Gandhakashala
16	Paramoola	20	Veliyan, Thondi, Chenellu, Mullanpunja, Mullankaima, Rakhtashali, Gandhakashala, Njavara
17	Manikeni	15	Veliyan, Thondi, Chenellu, Gandhakashala,
18	Mathamangalam	20	Adukkan, Thondi, Gandhakashala
19	Rampalli	8	Adukkan
20	Thakarpadi	8	Thondi, Adukkan
21	Kuttiyottil	22	Veliyan, Gandhakashala, Thondi, Chenellu, Okapunja
22	Madathuvayal	8	Veliyan
23	Puthoorvayal	8	Gandhakashala, Veliyan, Adukkan, Mullankaima
24	Kommayadu	8	Jeerakashala
25	Nambikolly	1	Adukkan, Jeerakashala
26	Anery	5	Chomala
27	Puliyarmala	3	Kalladiyaran
28	Tirunelli	5	Gandhakashala, Adukkan
29	Paralassery	2	Adukkan, Gandhakashala
30	Manamathamula	8	Adukkan
		309.25 acres	

(Source: Primary data collected from field as part of the study 2021)

8.4.1 Conservation of Traditional Rice Varieties and Preservation of Ecosystem Services

With the establishment of the RSVs, farmers were helped to maintain the on-farm conservation of 20 traditional rice varieties (*Mullankaima, ThonooramThondi, Karuvalicha,, Kunjootti Matta, Marathondi, Oonavattan, Chenthadi, Koduveliyan, Kalladiyaran, Valichoori, Chenellu, Gandhakashala, Chomala, Jeerakashala, Veliyan, Thondi, Kottathondi, Kaima, Mannuveliyan, and Adukkan*). The seed improvement activity focused on 10 selected specialty rice varieties, namely *Adukkan, Veliyan, Chenellu, Chenthadi, Chomala, Kalladiyaran, Thondi, Mullankaima, Jeerakashala, and Gandhakashala*. These are special in terms of their characteristic traits like fragrance, medicinal or nutritional properties, tolerance or resistance to the weather variations, early or late maturity, and so on. Until this intervention, no markets existed for the sale of seeds of TRVs. Through the initiative, farmers were given access to quality TRV seeds for cultivation in exchange for returning the seeds the year after the harvest.

In a span of 6 years from 2015 to 2021 around 27.13 Qtl (2.71 tonnes) of seeds of traditional rice varieties were exchanged through participatory seed exchanges across Kerala (12 districts) and Tamil Nadu (three districts) states of India.

After the RSV programme intervention, there has been an increase in TRV cultivation area in the seed villages from 248 acres to 309.25 acres, a growth of 24.7%, which indicates the interest of farming communities in the cultivation of traditional rice varieties.

8.4.2 Promotion of Climate-Resilient Traditional Varieties

Many traditional varieties have been found to be tolerant to flood and drought conditions by farming communities. Hence, the interest in traditional varieties as a climate change adaptation strategy is increasing slowly and steadily. Gopi and Manjula (2018) have noted that farmers reported many varieties tolerant to flood and drought conditions in the region. These TRVs were promoted as part of the RSV programme alongside climate-resilient agricultural practices. Promotion of the *Kalladiaran* variety as a promising drought-tolerant and short-duration crop helped the farmers in RSVs to cope with the aftermath of heavy rainfall during monsoons. They were also able to cover losses in other varieties if cultivated after the rainy season. In the aftermath of recurrent floods in 2018 and 2019, seed exchanges of these varieties helped communities to cope with the damage caused. After these floods, the majority of farmers lost their seeds, and certain released varieties like Athira, Uma, and Airangana could not survive the submerged flood conditions. In this regard, the linkage of seed villages across Wayanad aided in the seed exchange of resilient traditional rice varieties. The intervention enabled the seed exchange of

170.2 kg in 2018, which increased to 1279.35 kg (1.28 tonnes) in 2019 after the floods and 767.2 kg (0.76 tonnes) in 2020. Moreover, seeds were supplied to other farmers from different regions thinking ahead to any further unexpected weather events. The seed festivals conducted after the disaster events also gave farming communities outside the RSV intervention areas access to resilient TRVs. Overall, the intervention directly benefitted 340 farming families in the RSVs and indirectly benefitted more than 3000 farming families. The proven adaptive qualities of traditional varieties were demonstrated in their resilience during the recurrent floods and gave minimum price assurance on account of the maintained grain quality.

A major intervention towards climate adaptation and improved soil fertility was the establishment of pulse farms in existing RSVs during the off-season. Paddy fields are generally left fallow in the region after the harvest; therefore, cultivation of other crops in the off-season was promoted as part of the programme. Cultivation of vegetable crops and pulses was facilitated through distribution of seeds and planting materials for seasonal vegetables, pulse crops, and elephant foot yam. From 2018 onwards, nine pulse farms were established on 64 acres of land, helping to improve the soil fertility and nutrients through cultivation of pulse crops (traditional varieties of cowpea) after the harvest of paddy. On fields that were previously kept fallow after the harvesting of paddy, cowpea cultivation and incorporation of green manuring helped enhance soil fertility as legumes are known to fix nitrogen and enhance the fertility of soil. This intervention not only helped improve soil health, but the crops also served as a source of vegetables for the farm families and fodder for farm animals.

8.4.3 Socio-Economic Empowerment

Through RSVs, the informal seed exchange system was revived into a more structured network for participatory seed exchange to ensure conservation and cultivation of TRVs. RSVs were linked with a marketing channel through a Farmer Producer Organisation (FPO) named WAMPCo. to facilitate the sale of grains. Before linking the farming community with the market, it was essential to develop quality seeds. Ensuring and maintaining the purity of seeds is imperative to harness the benefits and special features of the traditional varieties. It is a major challenge to ensure adequate production and supply of pure seeds when promoting the large-scale cultivation of specialty varieties. Training of RSV farmers on quality seed production through pure line panicle selection, weeding and rouging of off-types, and efficient seed storage practices led to the production of good quality seeds of the TRVs. The supply of 88 metal seed storage boxes to the RSVs protected the seeds from disease and pests and allowed better post-harvest management. Good quality seeds produced through the intervention allowed for better market linkage in return. The RSV and seed festival interventions, coupled with similar initiatives taken by farmer-centred rice producer collectives, generated good awareness amongst consumers related to the nutritional qualities of traditional rice varieties.

With the intervention, linkage was facilitated between seed village farmers and markets, which eventually led farmers to fetch better prices compared to 24 INR/kg (about 0.30 USD) from the open market and smaller mills to 34 INR/kg (about 0.42 USD) from the Wayanad Rice Mill and certain FPOs interested in good quality seeds and grains of the *Veliyan, Adukkan, Thondi, Chennelu, Chomala, Kalladiaran,* and *Chenthadi* varieties.

Through the linkage, farmers were also able to fetch better prices for special aromatic rice varieties (*Gandhakashala, Mullankaima,* and *Jeerakashala*) that were promoted as part of the RSV intervention, not just from FPOs, but also from other potential buyers like Supply Co., other farmers, and private buyers. With the increased access to seeds of aromatic rice varieties, RSV farmers were able to cultivate these three varieties that sell at premium prices ranging from 70 to 110 INR/kg (0.86–1.35 USD). Due to increased consumer awareness, demand for these varieties gradually increased, resulting in premium prices roughly 40% higher than other varieties.

Furthermore, the networking of farmers with Seed Care, an association of traditional seed growers, also helped them to understand their rights and gain recognition amongst their peers. This helped communities to make strategic decisions regarding the quantity of seeds to be sold and other concerns regarding paddy cultivation. Developing a structured seed exchange network helped communities meet their livelihood needs in times of seasonal shortages, losses due to natural disasters, and other financial concerns.

In addition, as part of the seed festival programme, farmers were given recognition for their efforts in conservation of agro-biodiversity in the form of the "Community Agrobiodiversity Conservation Award" facilitated through the Wayanad Tribal Development Action Council (WTDAC), a grassroots organisation facilitated by MSSRF in the past. Over time, seed festivals have drawn wider attention to the value of traditional crop varieties. At every seed festival, two cash awards were rewarded to Adivasi farm families (indigenous community) from the region: one to the family that conserved the most native rice varieties (25,000 INR, or about 300 USD) and one to the family that conserved the most plant and animal breeds on farm (15,000 INR, or about 180 USD). These awards and recognition motivated other farmers and communities to make efforts to conserve agro-biodiversity. Accordingly, awareness on traditional varieties grew and created a higher demand amongst consumers, even at premium prices. As a result, the farming communities engaged in the cultivation of traditional varieties highly benefitted as they received better prices for their produce and recognition as well.

Fig. 8.5 Fish cultivation in empty paddy fields (Photo credit: Vipindas, P 2021)

8.4.4   Promotion of Alternate Livelihoods and Recreational Activities

Some other alternate uses of paddy fields during the fallow season were also promoted as part of the intervention to allow for the expansion of livelihood opportunities. Instead of keeping the land fallow, paddy fields were utilised as ponds for fish cultivation (Fig. 8.5), and for the cultivation of vegetables and pulses that provided alternate livelihood options as well as nutrition security. During summer months, certain paddy fields were made into football playgrounds for youngsters, helping to create recreational activities for enjoyment.

8.5   Discussion: Opportunities, Challenges, and the Way Forward

The REbA approach based on interventions in the RSVs facilitated better management of the paddy agro-ecosystem, which can be termed as a vital socio-ecological production landscape (SEPL) in the context of local sustainable development. Over the period of the intervention, various opportunities and challenges were revealed that affect the rice paddy SEPL.

Rice paddies as a type of SEPL hold immense importance on account of the diverse benefits they provide to communities. The opportunities identified throughout the intervention period which require further interventions include: value addition to TRVs, protection and enhancement of agro-biodiversity, creation of more employment opportunities in the paddy agro-ecosystem, and development of community-centric solutions to address food and nutrition security through management of the paddy agro-ecosystem. Major challenges observed throughout the programme period with respect to interventions and paddy cultivation were: increasing climate variability, expansion of cash crops, shortage of labour, insufficient incentives for farmers engaged in conservation, lack of community-centric agricultural innovations, and increased use of chemical fertilisers and pesticides. Future efforts need to be made to address these challenges by implementing development interventions in the concerned areas.

As evident from field observations and various reports, Wayanad has been experiencing various weather vagaries over the past few years, particularly the continuous occurrence of floods due to erratic rainfall during the monsoon season and drought conditions during the summers. Studies indicate the importance of implementing farm management practices based on agro-biodiversity and ecosystem services that provide adaptation benefits in light of climate change (van Noordwijk et al. 2018). Field interviews and observations showed how resilient some traditional varieties were to harsh weather conditions, such as droughts and floods. *Veliyan*, one of the traditional rice varieties, is especially worth mentioning as it has the particular traits of being tolerant to both flood and drought. Farmers also mentioned two different variants, namely *Chettuveliyan* (best for marshy land) and *Mannuveliyan* (best for sandy dry soil), for their climate-resilient qualities. Similar results have also been reported in a study done by Gopi and Manjula (2018) that mentioned a number of traditional varieties with climate resilience in Wayanad, i.e. flood-tolerant (*Adukkan, Veliyan, Chenthadi*) and drought-tolerant (*Kalladiaran, Veliyan, Thondi*) varieties. Further impetus is needed for field-based agronomical studies and lab-based research to delve deeper into the characteristics of resilient varieties and work to mainstream them in production systems.

It is evident that agro-ecosystems have undergone significant degradation on account of various anthropogenic activities, leading to the loss of agro-biodiversity. Wayanad, as stated above, has also undergone momentous changes in its agro-ecosystem due to land use changes, modern agriculture techniques, anthropogenic activities, and climate change. Hence, taking action for ecosystem restoration was quite necessary to retain or enhance the biodiversity of the paddy agro-ecosystem. The intervention through the RSVs was a major step towards conservation of agro-biodiversity and enhancement of livelihoods. Ecosystem restoration, as reported by Gann and Lamb (Eds., Gann and Lamb 2006), should improve biodiversity conservation, improve livelihoods, empower local people, and improve ecosystem productivity.

Through the establishment of RSVs, the traditional genetic diversity of rice paddies was conserved by improving access to and the availability of 10 traditional rice varieties, along with the increase in area under cultivation (from 248 to 309.25 acres) due also to the special characteristics of such varieties in relation to aroma, nutrition, and climate change. Moreover, through the intervention, improved access to desired seeds helped improve the livelihoods of the families engaged in traditional paddy cultivation. The intervention facilitated the linkage between farmers and potential buyers interested in the purchase of selected traditional varieties, thus enabling farmers to receive better remuneration for their produce. Furthermore, timely events like seed festivals and the training and capacity-building programmes also led to the socio-economic uplifting of the communities. As an outcome of the intervention governed with RSVs, local self-governance in the region, especially the Panchayati Raj Institutions initiated remunerative schemes for farmers that are conserving traditional rice varieties. Special subsidies were announced for farmers cultivating Jeerakashala and Gandhakashala varieties to encourage conservation of specialty rice varieties. Still, similar efforts are needed to encourage farmers to not only grow aromatic rice varieties but other traditional varieties as well for restoration of traditional paddy agro-ecosystem in long term. In a broader context, efforts in agro-biodiversity conservation and socio-economic empowerment of the farm families in the paddy SEPL of Wayanad are the primary stepping stones towards ecosystem restoration efforts in the future.

8.6 Conclusion

Through the integrated interventions, the REbA approach helped smallholder rice farming communities in Wayanad to improve their climate adaptation practices. The RSV programme resulted in stopping on-farm erosion of rice genetic diversity and created wider awareness of the nutritional and climate-resilient qualities of traditional rice varieties. RSVs as part of the intervention have taken up cultivation of atleast two to atmost eight traditional varieties and thus have led to conservation of the associated diversity of flora and fauna in rice fields. The increased consumer awareness on the therapeutic and nutritional value of some of the traditional rice varieties ensured better premium prices in the market and aided in the improvement of livelihoods in the farming communities. Awareness created amongst governance institutions, like the Panchayati Raj Institution (PRI) members, helped to increase allocation of funds as production subsidies, and special subsidies for medicinal and aromatic rice varieties. The farmers of the seed villages are able to sell their excess produce every year at a good rate, which provides them with better returns from the local market. The programme generated impacts in reviving the resilient, climate-friendly agricultural practices and varietal diversity owned by the traditional rice farming communities. Moreover, the indigenous knowledge associated with traditional paddy cultivation was also revived with the conservation of TRVs. The RSV programme gave an alternative solution to the problems associated with informal seed systems and declining rice diversity, making communities more self-reliant and self-sufficient through conservation, cultivation, marketing, and consumption of TRVs. Similar interventions are needed to pass on the agricultural heritage of indigenous communities merged with scientific knowledge to build resilient farming systems in the future. An enabling environment must be created that can serve the dual purpose of conservation of agro-ecosystems along with the development of sustainable livelihoods for farming communities.

Acknowledgements This study would not have been possible without the support of the Department of Science & Technology, Government of India, the invaluable cooperation of the farming communities, field professionals at MSSRF-CAbC, and other government and non-government personnel. We wholeheartedly acknowledge the support provided throughout the entire study period.

References

Chivenge P, Angeles O, Hadi B, Acuin C, Connor M, Stuart A, Puskur R, Johnson-Beebout S (2020) Ecosystem services in paddy rice systems. In: Rusinamhodzi L (ed) The role of ecosystem services in sustainable food systems. Academic Press, pp 181–201

Cox PA (2000) Will tribal knowledge survive the millennium? Science 287:44–45. https://doi.org/10.1126/science.287.5450.44

Deb D (2014) Folk rice varieties, traditional agricultural knowledge and food security. In: Alvarez C (ed) Multicultural knowledge and the university. Multiversity and Citizens International, Penang, pp 45–57

Easterling WE, Aggarwal PK, Batima P, Brander KM, Erda L, Howden SM, Kirilenko A, Morton J, Soussana JF, Schmidhuber J, Tubiello FN (2007) Food, fibre and forest products. In: Parry ML, Canziani OF, Palutikof JP, van der Linden PJ, Hanson CE (eds) Climate change 2007: impacts, adaptation and vulnerability. Cambridge University Press, Cambridge, pp 273–313

Eliazer Nelson ARL, Ravichandran K, Antony U (2019) The impact of the green revolution on indigenous crops of India. J Ethnic Foods 6(1):1–10

Food and Agriculture Organization (FAO) (2009) How to feed the world in 2050. Popul Dev Rev 35(4):837–839. https://www.fao.org/fileadmin/templates/wsfs/docs/expert_paper/How_to_Feed_the_World_in_2050.pdf

Fox TA, Rhemtulla JM, Ramankutty N, Lesk C, Coyle T, Kunhamu TK (2017) Agricultural land-use change in Kerala, India: perspectives from above and below the canopy. Agric Ecosyst Environ 245:1–10

Gann GD, Lamb D (eds) (2006) Ecological restoration: a means of conserving biodiversity and sustaining livelihoods. Society for Ecological Restoration International, Arizona

Godfray HCJ, Beddington JR, Crute IR, Haddad L, Lawrence D, Muir JF, Pretty J, Robinson S, Thomas SM, Toulmin C (2010) Food security: the challenge of feeding 9 billion people. Science 327(5967):812–818

Government of Kerala (GoK) (2011) Panchayath level statistics. Various Districts in Kerala, Department of Economics and Statistics, Government of Kerala, Thiruvananthapuram

Government of Kerala (GoK) (2019) Agricultural statistics 2019–20. Department of Economics & Statistics. Government of Kerala, Thiruvananthapuram, p 11p

Gopi G, Manjula M (2018) Speciality rice biodiversity of Kerala: need for incentivising conservation in the era of changing climate. Curr Sci 114:997–1006

Hari A, Kumar NK (2016) Scenario analysis of rice cultivation in Kerala. J Extens Educ 28(4):
5760–5763

Jose M, Raj RD, Vinitha MR, Madhu R, Varghese G, Bocianowski J, Yadav R, Patra BC, Singh
ON, Rana JC, Kurmari SL (2018) The prehistoric Indian Ayurvedic Rice Shashtika is an extant
early domesticate with a distinct selection history. Front Plant Sci 9:1203

KAU (2018) WayanadanNellinangal: directory of farmers conserving traditional rice varieties,
Govt of Kerala Dept of ag. Dev. and Farmers Welfare – Kerala Agricultural University

Kumar BM, Kunhamu TK (2021) Ecological and historical perspectives of rice cultivation in
Kerala: a synthesis. ORYZA-An Int J Rice 58(2):241–261

Kumar BM, Takeuchi K (2009) Agroforestry in the Western Ghats of peninsular India and the
satoyama landscapes of Japan: a comparison of two sustainable land use systems. Sustain Sci
4(2):215–232

Kumar NA, Lopus M, Raveendran T, Vipindas. (2022) Making Rice-farming system more climate
resilient and nutrition sensitive: heritage of Kurichiya tribe Community of Western Ghats. In:
Rakshit A, Chakraborty S, Parihar M, Singh Meena V, Kumar Mishra P, Bahadur Singh H (eds)
Innovation in small-farm agriculture. Taylor & Francis, pp 287–297

Kumar P (2017) Green revolution and its impact on environment. Int J Res Humanities Soc Sci 5(3):
54–57

Maffi L (2002) Endangered languages, endangered knowledge. Int Soc Sci J 54(173):385–393

Manilal KS (1991) Ethnobotany of the Rices of Malabar. In: Contribution to ethnobotany of India.
Scientific Publishers, Jodhpur, pp 243–253

Millennium Ecosystem Assessment (2005) Ecosystems and human Well-being: synthesis. Island
Press, Washington, DC

Parameswaran P, Kumar AN (2017) An account of the 'useful weeds' associated with wetland
paddy fields (Vayals) of Wayanad, Kerala, India. Ann Plant Sci 6(1):1516–1526

Parayil G (1992) The green revolution in India: a case study of technological change. Technol Cult
33(4):737–756

Richharia R, Govindasamy S (1990) Rices of India. Academy of Development Science, Karjat

Roy K, Mukherjee A, Maity A, Shubha K, Nag A (2018) Protecting non-basmati indigenous
aromatic rice varieties of West Bengal, India under geographical indication: a critical
consideration. In: Roy C (ed) The role of intellectual property rights in agriculture and allied
sciences. Apple Academic Press, pp 295–318

Salinger MJ, Sivakumar MVK, Motha R (2005) Reducing vulnerability of agriculture and forestry
to climate variability and change: workshop summary and recommendations. In: Salinger J,
Sivakumar MVK, Motha RP (eds) Increasing climate variability and change. Springer, Dordrecht, pp 341–362

Sathya A (2013) Are the Indian rice landraces a heritage of biodiversity to reminisce their past or to
reinvent for future? Asian Agri-History 17:221–232

Secretariat of the Convention on Biological Diversity (CBD) (2009) Connecting biodiversity and
climate change mitigation and adaptation: report of the second ad hoc technical expert group on
biodiversity and climate change, Montreal. Tech Series 41:126

Thakur AK, Uphoff NT (2017) How the system of rice intensification can contribute to climate-
smart agriculture. Agron J 109(4):1163–1182

Umadevi M, Pushpa R, Sampathkumar KP, Bhowmik D (2012) Rice-traditional medicinal plant in
India. J Pharmacogn Phytochem 1(1):6–12

van Noordwijk M, Duguma LA, Dewi S, Leimona B, Catacutan DC, Lusiana B, Öborn I, Hairiah K,
Minang PA (2018) SDG synergy between agriculture and forestry in the food, energy, water and
income nexus: reinventing agroforestry? Curr Opin Environ Sustain 34:33–42

Zeng Y, Zhang H, Wang L, Pu X, Du J, Yang S, Liu J (2010) Genotypic variation in element
concentrations in brown rice from Yunnan landraces in China. Environ Geochem Health 32(3):
165–177

The opinions expressed in this chapter are those of the author(s) and do not necessarily reflect the views of UNU-IAS, its Board of Directors, or the countries they represent.

Open Access This chapter is licenced under the terms of the Creative Commons Attribution-NonCommercial-ShareAlike 3.0 IGO licence (http://creativecommons.org/licenses/by-nc-sa/3.0/igo/), which permits any noncommercial use, sharing, adaptation, distribution and reproduction in any medium or format, as long as you give appropriate credit to UNU-IAS, provide a link to the Creative Commons licence and indicate if changes were made. If you remix, transform, or build upon this book or a part thereof, you must distribute your contributions under the same licence as the original. The use of the UNU-IAS name and logo, shall be subject to a separate written licence agreement between UNU-IAS and the user and is not authorised as part of this CC BY-NC-SA 3.0 IGO licence. Note that the link provided above includes additional terms and conditions of the licence.

The images or other third party material in this chapter are included in the chapter's Creative Commons licence, unless indicated otherwise in a credit line to the material. If material is not included in the chapter's Creative Commons licence and your intended use is not permitted by statutory regulation or exceeds the permitted use, you will need to obtain permission directly from the copyright holder.

Chapter 9
Community-Based Approach to Wetland Restoration: Case Study of the Songor Wetland, Ghana

Raymond Owusu-Achiaw and Yaw Osei-Owusu

9.1 Introduction

Tropical wetlands play an important role in biodiversity conservation. According to Raburu et al. (2012), tropical wetlands provide biodiversity and ecosystem services that are vital for the survival of a wide range of both flora and fauna. Additionally, they serve as important habitats for both aquatic and other wetland-dependent species (Quarto and Thiam 2018). Wetlands also support the health and productivity of seascapes and landscapes, especially mangroves. Furthermore, wetlands play an important, life-supporting role for countless coastal communities and local people, who depend on wetlands for life and livelihoods (Raburu et al. 2012). They are also recognised for their important role in mitigating climate change, sequestering up to five times more carbon than other terrestrial forest ecosystems, and also serving as living buffers against the forces of storms and waves that can otherwise devastate a coastline (Raburu et al. 2012).

The Songor Wetland is one of the coastal wetlands in Ghana that has been designated as a Ramsar site under the Ramsar Convention on Wetlands and is also listed as a biosphere under UNESCO's Man and Biosphere (MAB) Programme. It covers an area of 11,500 ha, extending 20 km along the coast and 8 km inland behind a narrow strip of sand dunes on which small communities are situated. The wetland acts as a habitat and breeding ground for several species of both economic and ecological importance. Thus it has multi-functional benefits relating to the ecosystem, the economy, and ecotourism (Ghana National MAB Committee 2009). However, studies have also identified threats to the wetland, including over-fishing, pollution, mangrove exploitation, invasive species, infrastructure development,

R. Owusu-Achiaw (✉) · Y. Osei-Owusu
Conservation Alliance International, Accra, Ghana
e-mail: rowusu-achiaw@conservealliance.org

© The Author(s) 2023
M. Nishi, S. M. Subramanian (eds.), *Ecosystem Restoration through Managing Socio-Ecological Production Landscapes and Seascapes (SEPLS)*, Satoyama Initiative Thematic Review, https://doi.org/10.1007/978-981-99-1292-6_9

and bushfires (Fianko and Dodd 2018). These threats are compounded by other challenges, such as poor perception of public ownership of resources, limited commitment to sustainable use, and weak enforcement of laws governing natural resources. Such threats and challenges to the sustainable management and conservation of the wetland indicate that its socio-cultural importance has virtually been lost to the fringe communities.

In 1999, the Government of Ghana developed the Songor Ramsar Site Management Plan under the Coastal Wetlands Management Project (CWMP) to address threats facing the Songor Wetland and to enhance its conservation status in Ghana. A key component of the management plan was the participation and inclusion of local communities in the ecological restoration (Ofori-Danson 1999). The inclusion of local communities in restoring the wetland was re-emphasised by the Ghana National Man and Biosphere (MAB) Committee (2009) as one of the surest ways of enhancing the wetland's ability to sustain biodiversity and provide ecosystem services to improve the socio-economic and environmental conditions of the local people. Various studies have revealed the critical role of community engagement and involvement in restoration practices, including planting of mangroves, agroforestry, and afforestation (Abdullah et al. 2014; Francis et al. 2015; Tomeo et al. 2018).

Communities on the fringe of the Songor Wetland have adopted various strategies to address threats to the wetland in order to improve the socio-ecological health of the landscape and enhance their well-being. The wetland serves as a strategic livelihood support system and economic lifeline for the communities through its fisheries, minerals, and ecotourism resources, as well as its good edaphic conditions for agricultural purposes (Fianko and Dodd 2018). Communal restoration activities have been undertaken in the wetland in partnership with civil society organisations, government entities, the private sector, and academia, amongst others, as recommended in the Songor Ramsar Site Management Plan. However, few or no comprehensive studies have been conducted to understand how communal approaches are being applied to restore the ecosystem of the wetland and which communal approaches are effective in addressing current and emerging threats. Likewise, studies were needed on the challenges faced by these communal approaches and lessons that can be learnt for improvement, scalability, and/or replicability.

Due to the ecological importance of the wetland and threats to it, Conservation Alliance International (CA) carried out a number of activities, including community education on conservation measures, to assist in community restoration efforts and evaluation of restoration efforts carried out within the fringe communities for lesson learning. Additionally, CA-led efforts to carry out several wetland restoration activities jointly with different stakeholders, including communities, civil society organisations, government entities, the private sector, and traditional authorities, as recommended by the Songor Ramsar Site Management Plan from 2015 to 2020. One key element of the CA-led restoration exercises was the integration of local and traditional knowledge and cultural values to facilitate the restoration of the Songor Wetland. At the end of the 5-year restoration period, CA also evaluated the different restoration approaches that had been adopted on all restored areas for replication, scalability, and lessons for socio-ecological production landscapes and seascapes (SEPLS) management.

To examine communal approaches to restoring the Songor Wetland landscape, CA conducted a study within the landscape. The main objective of this study was to gain a better understanding of communal approaches to wetland restoration and their role in enhancing the socio-ecological health of the Songor Wetland landscape of Ghana. Specifically, the study sought to:

(i) Assess the ecological value of the wetland for the communities.
(ii) Determine communal approaches/coordination to enhance the ecological restoration of the wetland.
(iii) Document current management practices undertaken within the communal restoration approaches.
(iv) Assess emerging challenges to community approaches to wetland restoration.

9.2 Methodology

9.2.1 Study Area

The Songor Wetland is located in the Greater Accra Region and was designated as a Ramsar site in 1992. It falls on the boundary of two flyways of water birds, the East Atlantic Flyway, and the Mediterranean Flyway. According to the 1971 Ramsar Convention on Wetlands of International Importance, wetlands are areas of land whose soil is saturated with moisture permanently or seasonally and are covered partially or completely with shallow pools which may be freshwater, saltwater, or brackish. The Songor Wetland has an abundance of all types, both freshwater, saltwater, and brackish wetlands (Ntiamoa-Baidu and Gordon 1991). The wetland protects some animal species listed in the IUCN Red List that are also protected in Ghana. It has the highest total tern count on the Ghana coast and supports naturally important populations (over 10% of the total coastal count) of at least 23 species of water birds (Piersma and Ntiamoa-Baidu 1995). The area is well known for its development, particularly involving tourism and salt production. Communities are also known to be very receptive to conservation projects (Ghana National MAB Committee 2009). According to Fianko and Dodd (2018), the Songor Wetland is divided into three main sections, namely the Core Area, Buffer Area, and Transition Area (Fig. 9.1).

The Core Area in the south is regarded as having the most important feeding and roosting sites for water birds, as well as nesting sites for marine turtles and fish species. It has two ecosystem types—marine and brackish lagoons. The Buffer Area is composed of the northern sections of the open lagoons and the surrounding floodplain, as well as some coastal land. Its ecosystem is also suitable for the breeding of fish, birds, and nesting turtles, and is thus very sensitive. The area is characterised by land use types such as farming, controlled salt mining, fishing,

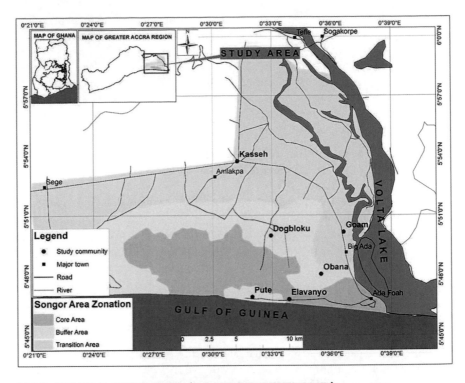

MAP OF PROJECT SITE (SONGOR WETLAND)

Fig. 9.1 Map showing the study site [Source: Ghana National MAB Committee (2009)]

Table 9.1 Basic information of the study area

Country	Ghana
Province	Greater Accra
District	Ada West
Size of geographical area (hectares)	28,978
Dominant ethnicity(ies), if appropriate	Dangmes, Ewes
Size of case study/project area (hectares)	11,500
Dominant ethnicity in the project area	Dangmes
Number of direct beneficiaries (people)	2000
Number of indirect beneficiaries (people)	44,000
Geographic coordinates (latitude, longitude)	5°48′30.9″N, 0°27′59.2″E

Source: Prepared by authors

tourism, and aquaculture on a minor scale. There is an enormous potential for the development of both aquaculture and salt production on an industrial scale. The Transition Area is mainly composed of settlements with a thin stretch of coastline on both ends. This area is fast developing. Current land uses include aquaculture, markets, schools, mechanised shops, and tourism. Additional information on the project area can be found in Table 9.1.

9.2.2   Study Population and Sampling Method

According to Ghana Statistical Service (2021), the total population of the Songor Wetland is about 152,498 people with a split number of males (49.9%) and females (50.1%). A simple random sampling technique was adopted to select respondents for the study. A sample size of 376 was derived from the random selection of respondents of which 146, 130, and 100 were from the Core, Buffer, and Transition Areas, respectively, based on their relative populations (Kirk 2011; Etikan and Bala 2017). The sample sizes were further divided to get an equal representation of gender, thus 188 respondents for both males and females were sampled.

9.2.3   Data Collection

Both quantitative and qualitative research approaches were employed. Likewise, both primary and secondary data constituted a useful set of data that defined the state of restoration activities and their impacts on the socio-ecological status of the Songor Wetland. Farmer surveys, community consultations, focus group discussions (FGDs), district-wide stakeholder consultations, and key informant interviews constituted the main strategies to generate primary data. Questionnaires for surveying farmers were pre-tested using a selected group of respondents outside the target population to ensure the reliability and validity of the questions and responses.

Secondary data were collected from various sources including public records, institutional documents, management plans, policy statements, official publications, and other research works. Reviews of government documents on the Songor Ramsar Site Management Plan under the Coastal Wetlands Management Project (CWMP), including operational manuals and monitoring reports, constituted another form of data collection.

A participatory approach through FGDs was used to obtain information on the different communal restoration models that were implemented to improve degraded portions of the wetland. Two FGDs were conducted between 2020 and 2021. A community-based evaluation approach was used to evaluate the success of the different communal approaches based on each approach's effectiveness in responding to a set of restoration criteria (e.g. mangrove restoration, woodlot planting, control of invasive species, amongst others). Approaches were scored according to the following: 1—Very Unsatisfactory, 2—Unsatisfactory, 3—Weak Satisfactory, 4—Satisfactory, and 5—Very Satisfactory. Two evaluations were conducted between 2020 and 2021. The 27 representatives of relevant stakeholder groups that participated in the FGDs and the evaluations are shown in Table 9.2.

Two series of interviews were also conducted with the Wildlife Division (WD), Environmental Protection Agency (EPA), the traditional authorities, companies, and civil society groups, amongst others, to assess the threats to the Songor Wetland and efforts to address them through restoration exercises. Additionally, a Community Ecosystem Service Value Typology (CESVT) was used to assess how fringe communities value the wetland's ecosystem services. This information provided an insight into how this perceived value is likely to motivate certain community behaviours and facilitate the design of appropriate biodiversity conservation projects.

9.2.4 Data Analysis

The data collected were coded and analysis was carried out using Microsoft Excel and Statistical Package for Social Sciences (SPSS) version 27. The statistical outputs generated were summarised in tables for simplicity of interpretation. The analysis was done using descriptive statistics where issues of similarity and dissimilarity of responses were compared. The descriptive statistical analysis facilitated the quantitative comparison of the responses.

Table 9.2 Selected key persons for FGDs and community evaluations

Interest group	Collective stake	Entities	Relevance to the Songor Wetland	Key representative[a]	Number of representatives
Government	Leads/manages government-led development initiatives and activities inside and outside the Songor Wetland	Wildlife Division	Responsible for management of the wetland	One Wildlife Manager for the project area	1
		Environmental Protection Agency (EPA)	Environmental monitoring of the wetland	One Regional Director	1
		Minerals Commission	Monitoring of impact of mining on the wetland	One Regional Manager	1
		District Assemblies	Physical and development planning for the entire project landscape	One District Planning Officer of the project area	1
		Ghana Tourism Authority	Facilitating investment and development of potential tourism sites of the wetland	One Regional Manager	1
		Water Resources Commission	Monitoring the health of water bodies in the project area	One Regional Manager	1
Traditional authorities	Land ownership by occupancy and defence of the land	Chiefs (target project communities)	The traditional overlord of target communities that depend directly and indirectly on Songor Wetland resources	Six Community Heads with one chief with jurisdiction over a community	6
Private sector	Major players in private sector-led initiatives within the landscape	Tourism operators	Farmer education and sensitisation on sustainable farming practices	Six Local Managers of major hotels	6
		Fish processors/commercial fishing companies	Large-scale fishing, fish farming, and processing of fishery resources	Two Directors of major operating companies	2
		Mining companies	Salt mining activities		2

(continued)

Table 9.2 (continued)

Interest group	Collective stake	Entities	Relevance to the Songor Wetland	Key representative[a]	Number of representatives
				One Manager of a major salt mining company in the landscape	
Civil society (NGOs/CBOs)	Major players in civil society and donor initiatives and activities within the project area	NGOs	Extensive information on the wetland landscape which could influence decision-making at the local level; Promotion of green alternative livelihoods and restoration activities within the landscape	Three Managers at the local office in the project area	3
		Ada Stakeholders Association (CBO)	A locally based civil society with links to the traditional authority, and in possession of information on local conditions of the wetland	One Director at the local office in the project area	1
Research and academia	Major scientific/knowledge body within the landscape	Man & Biosphere (MAB) Committee	Extensive scientific knowledge of the project area	One Director at the office in the project area	1
Total					27

Source: Prepared by authors

Note: [a]The individuals constituting each stakeholder group are managers and top officials who have been mandated to speak on issues with respect to their interest within Songor Wetland

9.3 Results

9.3.1 Demographic Characteristics of Respondents

Statistics on key demographic characteristics that were generated during the study included the age, level of education, and occupation of respondents (Table 9.3). More than half (55.5%) of the respondents were male, arising from the dominance of males in the use of the Songor Wetland (Table 9.2). Additionally, 80% of the respondents were below the age of 50, and 63.9% practised agriculture as their main occupation. Education plays a significant role in the acquisition and use of science and new technologies to undertake restoration activities. More than 50% of respondents had no formal education. As such, concerted efforts on awareness raising and capacity building to enhance restoration activities will be required.

9.3.2 Assessment of the Ecological Value of the Songor Wetland

As noted above, a CESVT approach was adopted to determine the perceived value communities place on the Songor Wetland ecosystem (Table 9.4). Most respondents put a higher premium on provisioning services, accounting for 53% of responses on the ecosystem services delivered by the Songor Wetland. This result shows how communities rely mostly on these services (provisioning) for their daily sustenance or livelihoods. Provisioning services provided by the wetland include fishes, oysters, fuelwood, and crabs, amongst others. Next was supporting services (31%) that are necessary to produce all other ecosystem services, including soil formation and

Table 9.3 Demographic characteristics of respondents

Demographic characteristics		Frequency ($n = 376$)	Percentage
Gender	Male	209	55.6
	Female	167	44.4
Age group (years)	20–29	105	28
	30–39	120	32
	40–49	75	20
	50–59	38	10
	60–69	30	8
	70 and above	8	2
Education	No education	195	52
	Basic education	143	38
	Higher education	38	10
Occupation	Agriculture	240	63.9
	Service	76	20.2
	Industry	60	15.9

Source: Prepared by authors based on field survey data

Table 9.4 Perceived ecosystem value based on CESVT

Ecosystem services	Traditional authorities	Farmers	Fishermen	Fishmongers	Men's groups	Women's groups	Youth	Total	Total (%)
Provisioning	26	32	39	35	31	21	24	199	53
Regulating	8	6	5	6	6	4	6	41	11
Cultural	6	3	3	2	3	2	0	36	5
Supporting	13	22	24	17	14	13	14	117	31
Total	53	63	71	60	53	40	44	376	100

Source: Prepared by authors based on field survey data

Fig. 9.2 Crabs and oysters are amongst provisioning services derived from the wetland (Source: Field picture taken by authors. Photo credit: Raymond Owusu-Achiaw)

retention, water cycling, nutrient cycling, primary production, production of atmospheric oxygen, and provision of habitat for biodiversity. Some regulating services provided by the wetland include erosion control, carbon sinks, and climate regulation. Cultural services include aesthetic inspiration that provides opportunities for tourism and recreation as well as spiritual affinity for the sea turtles.

There were observed disparities in values across and within each ecosystem service perceived by the different community groups. The main driver of these disparities was how each ecosystem service was contributing to livelihood sustenance. Farmers, fishermen, and fishmongers, as well as men's groups, found the most value in provisioning services, which have the most direct impact on their livelihoods (Fig. 9.2). Sustenance of livelihoods is essential no matter the sources of livelihoods. This explains why community members are likely to buy into ecosystem restoration activities provided restoration improves supporting services that enhance their livelihoods. Regulating services were mostly relegated by community members as less valuable in comparison to provisioning and supporting services. Discussions through community meetings revealed that elements of regulating services' value were too technical and scientific for the community members to fully understand and appreciate. Furthermore, the traditional authorities, as the custodians of the cultural integrity of the landscape, attributed a comparatively higher value to cultural services compared to the other community groups. The mostly intangible nature of cultural services was revealed to be one of the reasons for the low value attributed by community members. The study revealed that the socio-cultural value of the landscape is gradually being lost to the youth for multiple reasons, including foreign cultural influence, reduced socio-cultural education, and lack of proper integration of youth into the cultural discourse.

9.3.3   Communal Approaches to Ecological Restoration of the Wetland

The study revealed that communities use diverse approaches and strategies to undertake restoration activities within the Songor Wetland (Fig. 9.3). The

Fig. 9.3 Level of approach/
coordination to wetland
restoration (Source:
Prepared by authors)

**Level of approach/coordination to wetland
restoration**

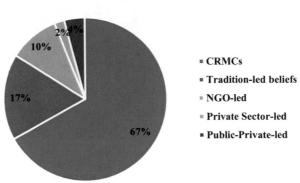

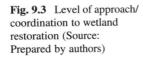

Community Resource Management Committee (CRMC) model accounted for about 67% of approach and coordination strategies for wetland restoration by the communities, followed by tradition-led beliefs in a distant second (17%).

The CRMC model has been very successful for most restoration activities because it is hinged on the concept of inter-agency approach/coordination (a multi-stakeholder approach). Inter-agency coordination is coordination that occurs between government agencies, private sector operators, non-governmental organisations, and regional and international organisations for the purpose of accomplishing a common objective. This approach has allowed for all stakeholders to make concerted efforts in a coordinated manner to restore the Songor Wetland, whilst also allowing for mainstreaming of restoration lessons at the local level through the district assembly. The conceptual framework of the CRMC model is shown in Fig. 9.4.

Fig. 9.4 Conceptual
coordination framework for
the CRMC restoration
model (Source: Prepared by
authors)

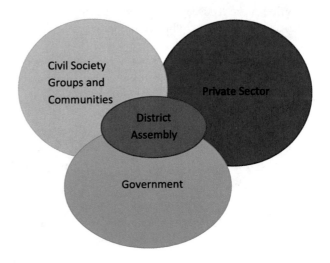

9.3.4   Current Management Practices Under the Communal Restoration Approaches

Current management practices carried out through communal approaches are aimed at biodiversity conservation, recovery of species, improvement of community livelihoods, and also enhancement of the socio-cultural value of the wetland. Restoration exercises included mangrove restoration, woodlot establishment, clearing of aquatic weeds, location of disoriented turtles, pest control, and data collection for monitoring (Fig. 9.5). These management practices were intended to prevent, halt, and reverse the degradation of the wetland in order to sustain the biodiversity and ecosystem services enjoyed by the local communities. Thus, ecosystem integrity and the sustainable development of the Songor Wetland and its fringe communities were expected to be enhanced through the adoption of these management practices. Additionally, alternate fuelwood was also provided to community members to reduce mangrove exploitation. Most of the restoration activities focused on mangrove restoration (55%), since unsustainable mangrove exploitation is the single most significant threat to the survival of the wetland. The next most common restoration activity was woodlot establishment (20%), as communities need alternate fuelwood to reduce their dependence on mangroves for wood products, including fuel. The mangrove ecosystem was restored with species such as *Rhizophora mangle* (red mangrove) and *Laguncularia racemosa* (white mangrove), which are native to the Songor Wetland, also aimed at enhancing the natural habitats of some biological elements (e.g. spawning fishes and sea turtles)(Fig. 9.6). Woodlots were established with *Acacia mangium* (mangium), which is fast growing and a soil nitrogen fixer, to produce an alternative source of fuelwood. Restoration activities such as clearing of aquatic weeds, location of disoriented turtles, pest control, and data collection for

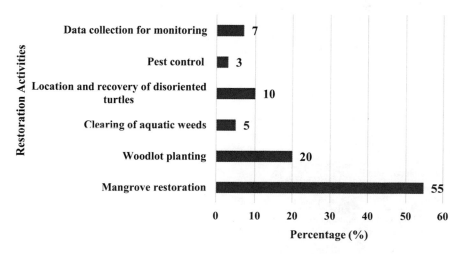

Fig. 9.5 Examples of restoration activities (Source: Prepared by authors)

Fig. 9.6 Red mangrove and *Acacia mangium* planting by community members (Source: Field pictures taken by authors; Photo credit: Raymond Owusu-Achiaw)

monitoring complemented the main restoration activities (e.g. mangroves and wood-lots) and constituted the smallest number of restoration activities undertaken under the management practices.

The study also evaluated the response of each communal restoration and the types of restoration activities undertaken within the project area to establish a success matrix for each approach. Responses from participants during the FGDs and evaluations were measured on a Likert scale and ranked in the following order: 1—Very Unsatisfactory, 2—Unsatisfactory, 3—Weak Satisfactory, 4—Satisfactory, 5—Very Satisfactory. A total score was aggregated for each communal restoration approach, as shown in Table 9.5. The CRMC model was found to be the most successful and satisfactory approach in restoring the wetland ecosystem. Its near perfect score is attributable to the involvement of all relevant actors in the restoration activities. This approach offered the most efficient mobilisation of resources (including finance), a sense of local community ownership, alignment of diverse interests,

Table 9.5 Success matrix of communal restoration approaches

Restoration activity ↓	Restoration approach					
	CRMC	Public-private-led	Tradition beliefs-led	Private sector-led	NGO-led	Score
Mangrove restoration	5	3	3	2	5	18
Woodlot planting	5	2	3	2	4	16
Clearing of aquatic weeds	4	3	2	4	3	16
Pest control	4	3	2	3	2	14
Location and recovery of sea turtles	5	2	3	2	4	16
Data collection and monitoring	4	2	2	1	3	12
Total	27	15	15	14	21	95

Source: Prepared by authors based on field survey data
Note: 1—Very unsatisfactory, 2—Unsatisfactory, 3—Weak satisfactory, 4—Satisfactory, 5—Very satisfactory

defined roles, and clear ecosystem sustainability pathways. The NGO-led approach was the second most successful and involved full community participation. One drawback was that most restoration activities were externally driven by donor funding, making sustainability an issue once funding is exhausted. The private sector-led and public-private-led approaches were deemed to be more profit-oriented in scope and lacked the full participation of community members, though funding was not a problem. Thus, sustainability through private sector- and public-private-led approaches was very low as compared to a multistakeholder approach (e.g. CRMC).

9.4 Discussion and Lessons Learnt

9.4.1 Discussion

Demographic information is an essential element in describing the characteristics of the study participants. The dominant age groups of the respondents were below 50 years. These ages indicate that the populace is more youthful, and that the bulk of persons involved were engaged in productive activities. Weiss and Börsch-Supan (2011) established that worker productivity ages amongst workers fall within the 20–50 years age bracket. Furthermore, the study revealed that 90% of the respondents had either no education or basic education. This could be one of multiple reasons that explain the weak appreciation for the value of the ecosystem goods and services provided by the Songor Wetland, as well as the many threats to it. Studies have shown that it is difficult for new innovations in conservation to be adopted by persons with only basic or no formal education (OECD 2016; Ntshangase et al. 2018). The level and quality of education tend to influence the rate of adoption of new conservation and natural resource management techniques (Brewer 2006; Coulibaly-Lingani et al. 2009). The findings of the present study showed that most community members are engaged in agriculture, which is consistent with the 2021 Ghana Population and Housing Census report, which indicated that most communities rely on fishery resources and good soil conditions in the Songor Wetland for agricultural production (Ghana Statistical Service 2021). This study also confirmed earlier studies, which indicate that the livelihoods of people within natural resource-rich areas of Ghana are predicated on subsistence agriculture (Appiah 2009; Ghana Statistical Service 2021).

The CESVT approach adopted to determine the value communities place on the Songor Wetland indicated that most respondents place a higher value on the provisioning and supporting services provided by the wetland. This shows that these services are crucial to the local livelihoods. Locally, the fringe communities depend on the wetland for food, raw materials, and firewood, amongst others (Lamptey and Ofori-Danson 2014). Additionally, some researchers have argued that an in-depth understanding of how people value ecosystem services is crucial for policymakers to conceptualise the contribution of ecosystems to human society (Yang et al. 2018).

The CRMCs provided inter-agency coordination that was perceived to be the best option for landscape projects because it provides the opportunity for different actors to act in a concerted and coordinated manner to achieve set goals and targets. It also empowers government agencies to mainstream the projects into their development plans and policies both at the local and national levels. Local governments in Ghana play a key role in administration and development of local areas and associated resources. The 1992 Constitution of the Republic of Ghana, in Article 240, therefore tasks the local government authorities (Metropolitan, Municipal, and District Assemblies—MMDAs) with the planning, initiation, coordination, management, and execution of policies with respect to all matters affecting the people within their areas of jurisdiction (ILGS 2016). This means that policy planning and mainstreaming of development processes at the national level start with the local government. Thus, the CRMCs offer a model for integrating lessons learnt from communal restoration activities into development policies.

Threats to the Songor Wetland that have negative impacts on livelihoods of community members are the main drivers that motivate restoration activities within the wetland. The main restoration activities undertaken by communities to address these threats include mangrove planting and establishment of woodlots to serve as an alternative source of fuelwood to ease mangrove exploitation for fuelwood. According to Lamptey and Ofori-Danson (2014), the vigorous pursuit of simultaneous mangrove restoration and woodlot planting is likely to address more than half of the challenges facing the Songor Wetland. This is corroborated by several studies, including Abdullah et al. (2014), Francis et al. (2015), and Newaz and Rahman (2019), which revealed that community-led wetland restoration activities not only ensure the socio-ecological health of the entire wetland ecosystem, but also empower communities to take ownership of and participate in the sustainable management of their own resources.

The Songor Wetland restoration initiatives by communities also face emerging challenges due to an increasing population and the onset of urbanisation. Changing land use types, such as mining and tourism (e.g. mainly driven by the improved natural ecosystem through restoration efforts), are having the greatest impacts on restoration activities and are mostly emanating from the buffer and transition areas of the wetland. This is made worse by the invasiveness of aquatic weeds (water hyacinth), the use of agrochemicals for agricultural purposes resulting in pollution, and the erosion of coastal areas due to more tidal waves than normal. According to the Ghana National MAB Committee (2009), human impacts on the Songor wetland ecosystem are expected to increase if proper land use management practices are not put in place and implemented. For the local community, the greatest benefits are increased employment opportunities from the mining and tourism industries.

Thus, the CRMC model offers an opportunity to continually address these emerging challenges because it enhances multi-stakeholder collaboration and coordination, local community ownership, and the sustainability of restoration activities, which ultimately improve the socio-ecological health of the wetland ecosystem. The expansion of the CRMC model into the buffer and transition areas of the wetland will go a long way to further addressing these emerging threats.

9.4.2   *Emerging Challenges to Community Approaches to Wetland Restoration*

The study explored some emerging challenges that communal restoration activities have faced in the past decade that negatively affect ongoing efforts to enhance the socio-ecological health of the Songor Wetland (Table 9.6). The findings revealed that the most impacting activities were mining (30%) and tourism (24%). Others included aquatic weeds, water hyacinth (19%), use of agrochemicals (16%), and soil erosion along the coast (11%), exacerbated by impacts of the COVID-19 pandemic.

Salt mining constitutes the main mining activity within the Songor Wetland that has emerged as a risk to restoration efforts. Salt mining is carried out all year round, and three million tonnes are expected to be produced annually from 2023. The use of heavy drilling and dredging machinery, as well as the removal of vast areas of new and existing mangroves for salt processing facilities, have affected the ecological health of the wetland. According to Ekrami et al. (2021) and Gbogbo (2007), salt mining imperils the health of wetland ecosystems and the survival of biodiversity.

Restoration efforts, especially planting of mangroves and removal of water hyacinth, enhanced the ecosystem of the wetland, which attracted new hoteliers to the burgeoning tourism industry (Fig. 9.7). This led to massive infrastructure development and other pressures from increased tourism affecting the health of the

Table 9.6   Challenges to restoration

Challenges	Frequency	Percent (%)
Mining	113	30
Tourism	90	24
Aquatic weeds	68	18
Use of agrochemicals	60	16
Soil erosion along the coast	45	12
Total	376	100

Source: Prepared by authors based on field survey data

Fig. 9.7 Water hyacinth that poses a threat from upstream to restoration exercises downstream being removed by a hotel staff person (Source: Field pictures taken by authors, Photo credit: Raymond Owusu-Achiaw)

ecosystem due to the habitat degradation of some key species within the wetland (e.g. marine turtles and species that thrive in the brackish ecosystem). Similar findings have revealed that the potential negative impacts of tourism on biodiversity often interact with wider landscape-level impacts such as land use changes that compromise biodiversity conservation (Ouboter et al. 2021; Newsome and Hughes 2016).

9.4.3 Lessons Learnt

This case study has provided important lessons from an examination of the implementation of Songor Wetland restoration activities. It gave perspectives on different approaches to communal restoration exercises. The tradition-led approach showed the importance of culture and spiritual affinity for certain biological elements (e.g. marine turtles) in motivating community members to restore the degraded wetland ecosystems. Similarly, private sector-led restoration efforts were primarily activities aimed at improving business patronage and revenues (e.g. removal of aquatic weeds). This shows that restoration activities were conducted based on interest and perceived benefits derived from ecosystem services by each stakeholder operating within the landscape. The biggest lesson is, therefore, that for restoration activities to be truly successful, the interest of diverse stakeholders ought to be considered in the restoration planning processes. This is where the CRMC model comes in handy as it takes into account the diverse stakes and interests of stakeholders during ecosystem restoration.

Evidence from the study clearly showed that livelihood improvement is a major driver that stimulates restoration activities by stakeholders, especially communities fringing the Songor Wetland. Therefore, the success of ecosystem restoration must be linked to the livelihood sustenance of community members for them to fully embrace such restoration activities. If this is not done, the success of ecosystem restoration will be limited.

Additionally, a multi-stakeholder approach to restoration (e.g. involvement of local government, private companies, scientists, communities, and civil society) is essential for resource mobilisation, community education, and capacity building, amongst others, to enhance the success of ecosystem restoration. This is evidenced in the CRMC model, which proved to be the most effective approach in the Songor Wetland ecosystem because it clearly exhibited the characteristics of a multi-sectoral restoration method.

Another key lesson is that progress in ecosystem restoration, especially for mangroves, contributes immensely to the ecological health of the wetland. In Songor, this has enhanced the ecotourism industry of the area, but also created residual impacts on the ecology of the wetland due to the development of ecotourism infrastructure and increased pressure from tourists. Therefore, restoration planning must include risk mitigation measures to ensure that improvements to livelihoods and the socio-economic situation of communities do not in turn impact negatively on the ecological health of a landscape or seascape.

9.5 Conclusion and Recommendations

This study has established that the provisioning and supporting services derived from the Songor Wetland are the driving factors that have stimulated interest in restoration of the ecosystem by local communities to address threats to the wetland. Furthermore, the CRMC model of restoring the wetland was found to be the most effective approach since it brings all relevant actors together to leverage technical, financial, and management resources to ensure that restoration activities are undertaken in a more coordinated and inclusive manner, thereby maximising on the positive impacts of restoration of the wetlands.

1. Based on this study, the following recommendations can be made. Through its statutory agencies responsible for the management of the Songor Wetland, the Ghanaian government should replicate and/or scale up the CRMC model to further improve the socio-economic and ecological sustainability of the Songor Wetland and other wetlands in the country.
2. Policymakers should support traditional authorities and the CRMCs to deepen local participation in the management and restoration activities by mainstreaming CRMCs into local government development plans to enhance the ecological health of the Songor Wetland and address the current and emerging challenges facing restoration efforts for wetlands in the country.

References

Abdullah K, Said AM, Omar D (2014) Community-based conservation in managing mangrove rehabilitation in Perak and Selangor. Procedia Soc Behav Sci 153:121–131. https://doi.org/10.1016/j.sbspro.2014.10.047

Appiah D (2009) Personifying sustainable rural livelihoods in forest fringe communities in Ghana: a historic rhetoric? J Food Agric Environ 7:873–877. https://www.academia.edu/53213357/Personifying_sustainable_rural_livelihoods_in_forest_fringe_communities_in_Ghana_A_historic_rhetoric. Accessed 12 Mar 2022

Brewer C (2006) Translating data into meaning: education in conservation biology. Conserv Biol 20(3):689–691. https://www.jstor.org/stable/3879232

Coulibaly-Lingani P, Tigabu M, Savadogo P, Oden PC, Ouadba JM (2009) Determinants of access to forest products in southern Burkina Faso. For Policy Econ 11(7):516–524. https://www.jstor.org/stable/24310769

Ekrami J, Nemati Mansour S, Mosaferi M, Yamini Y (2021) Environmental impact assessment of salt harvesting from the salt lakes. J Environ Health Sci Eng 19(1):365–377. https://doi.org/10.1007/s40201-020-00609-2

Etikan I, Bala K (2017) Sampling and sampling methods. Biom Biostat Int J 5(6):215–217. https://doi.org/10.15406/bbij.2017.05.00149

Fianko JR, Dodd HS (2018) Sustainable management of wetlands: a case study of the Songor Ramsar and UNESCO man and biosphere reserve in Ghana. J Wetl Environ Manag 6(1):45. https://doi.org/10.20527/jwem.v6i1.173

Francis R, Weston P, Birch J (2015) The social, environmental, and economic benefits of Farmer Managed Natural Regeneration (FMNR). World Vision. http://fmnrhub.com.au/wp-content/uploads/2015/04/Francis-Weston-Birch-2015-FMNR-Study.pdf. Accessed 2 Oct 2021

Gbogbo F (2007) Impact of commercial salt production on wetland quality and waterbirds on coastal lagoons in Ghana. Ostrich 78:87. https://doi.org/10.2989/OSTRICH.2007.78.1.12.56

Ghana National MAB Committee (2009) Ecological mapping of the Songor Ramsar site. Final report submitted to Environmental Protection Agency (EPA), Accra

Ghana Statistical Service (2021) Ghana 2021 population and housing census. General report. Ghana Statistical Service, Accra. https://census2021.statsghana.gov.gh/. Accessed 1 Sept 2021

ILGS (2016) A guide to district assemblies in Ghana, 2nd edn. Institute of Local Government Studies and Friedrich Ebert Stiftung Ghana. https://www.kassenanankanama.org/documents/local/DISTRICT%20ASSEMBLY_2nd%20Edition.pdf. Accessed 12 Jan 2022

Kirk RE (2011) Simple random sample. In: Lovric M (ed) International encyclopedia of statistical science. Springer, Berlin, pp 1328–1330. https://doi.org/10.1007/978-3-642-04898-2_518

Lamptey AM, Ofori-Danson PK (2014) Review of the distribution of waterbirds in two tropical coastal Ramsar Lagoons in Ghana, West Africa. West Afr J Appl Ecol 22(1):77–91. https://www.ajol.info/index.php/wajae/article/view/108001

Newaz MW, Rahman S (2019) Wetland resource governance in Bangladesh: an analysis of community-based co-management approach. Environ Dev 32(4):100446. https://doi.org/10.1016/j.envdev.2019.06.001

Newsome D, Hughes M (2016) Understanding the impacts of ecotourism on biodiversity: a multi-scale, cumulative issue influenced by perceptions and politics. In: Geneletti D (ed) Handbook on biodiversity and ecosystem services in impact assessment. Edward Elgar Publishing, pp 276–298

Ntiamoa-Baidu Y, Gordon C (1991) Coastal wetlands management plans, Ghana document prepared for the World Bank/EPA, Ghana. https://documents1.worldbank.org/curated/en/396621468251981053/pdf/511660ESW0Whit10Box342025B01PUBLIC1.pdf.   Accessed 25 Sept 2021

Ntshangase NL, Muroyiwa B, Sibanda M (2018) Farmers' perceptions and factors influencing the adoption of no-till conservation agriculture by small-scale farmers in Zashuke, KwaZulu-Natal Province. Sustainability 10(2):555. https://doi.org/10.3390/su10020555

OECD (2016) Innovating education and educating for innovation: the power of digital technologies and skills. OECD Publishing, Paris. https://doi.org/10.1787/9789264265097-en. Accessed 3 July 2022

Ofori-Danson PK (1999) Coastal Wetlands Management Project (CWMP): Songor Ramsar site management plan. Prepared for the Wildlife Division of the Forestry Commission. https://rsis.ramsar.org/RISapp/files/21563577/documents/GH566_mgt1506.pdf. Accessed 23 Dec 2021

Ouboter DA, Kadosoe VS, Ouboter PE (2021) Impact of ecotourism on abundance, diversity and activity patterns of medium-large terrestrial mammals at Brownsberg Nature Park, Suriname. PLoS One 16(6):e0250390. https://doi.org/10.1371/journal.pone.0250390

Piersma T, Ntiamoa-Baidu Y (1995) Waterbird ecology and the management of coastal wetlands in Ghana. Netherlands Institute for Sea Research & Ghana Wildlife Society for Ghana Coastal Wetlands Management Project

Quarto A, Thiam I (2018) Community-Based Ecological Mangrove Restoration (CBEMR): re-establishing a more biodiverse and resilient coastal ecosystem with community participation. Nat Faun J 32(1):39–45. https://www.fao.org/3/I9937EN/i9937en.pdf. Accessed 29 Jan 2022

Raburu PO, Okeyo-Owuor JB, Kwena F (2012) Community based approach to the management of Nyando Wetland, Lake Victoria Basin, Kenya. https://www1.undp.org/content/dam/kenya/docs/energy_and_environment/Nyando%20Book%20-%20FINAL%20MOST-internet.pdf. Accessed 12 Oct 2022

Tomeo T, Mahamoudou S, Averna EH (2018) From pledges to action: Africa's 100 million hectare restoration goal comes into focus. Nat Faun J 32(1):27–32. https://www.fao.org/3/I9937EN/i9937en.pdf. Accessed 2 Jan 2022

Weiss M, Börsch-Supan A (2011) Productivity and age: evidence from work teams at the assembly line. VfS annual conference 2011 (Frankfurt, Main): the order of the world economy—lessons from the crisis 48719. Verein für Socialpolitik/German Economic Association. https://ideas.repec.org/p/zbw/vfsc11/48719.html. Accessed 6 July 2022

Yang Y-CE, Passarelli S, Lovell R, Ringler C (2018) Gendered perspectives of ecosystem services: a systematic review. Ecosyst Serv 31(4):58–67. https://doi.org/10.1016/j.ecoser.2018.03.015

The opinions expressed in this chapter are those of the author(s) and do not necessarily reflect the views of UNU-IAS, its Board of Directors, or the countries they represent.

Open Access This chapter is licenced under the terms of the Creative Commons Attribution-NonCommercial-ShareAlike 3.0 IGO licence (http://creativecommons.org/licenses/by-nc-sa/3.0/igo/), which permits any noncommercial use, sharing, adaptation, distribution and reproduction in any medium or format, as long as you give appropriate credit to UNU-IAS, provide a link to the Creative Commons licence and indicate if changes were made. If you remix, transform, or build upon this book or a part thereof, you must distribute your contributions under the same licence as the original. The use of the UNU-IAS name and logo, shall be subject to a separate written licence agreement between UNU-IAS and the user and is not authorised as part of this CC BY-NC-SA 3.0 IGO licence. Note that the link provided above includes additional terms and conditions of the licence.

The images or other third party material in this chapter are included in the chapter's Creative Commons licence, unless indicated otherwise in a credit line to the material. If material is not included in the chapter's Creative Commons licence and your intended use is not permitted by statutory regulation or exceeds the permitted use, you will need to obtain permission directly from the copyright holder.

Chapter 10
An Integrated Seascape Approach to Revitalise Ecosystems and Livelihoods in Shimoni-Vanga, Kenya

Tamara Tschentscher, Nancy Chege, and Hugo Remaury

10.1 Introduction

Some of the world's most valuable coastal and marine resources, such as mangrove forests, seagrass beds, coral reefs, and deltas, are found on the Kenya Coast and support local livelihood and economic activities. However, the coastal and marine environment faces numerous issues and challenges. High rates of population growth, urbanisation, expansion of industrial developments, overexploitation of natural resources, and climate change, are among the key drivers of environmental degradation in the coastal region (NEMA 2017).

Located in the southern coastal region of Kenya, the Shimoni-Vanga Seascape faces a myriad of challenges despite its significance for the biodiversity maintained within its diverse ecosystems—such as mangrove forests, seagrass beds, and coral reefs—which support coastal communities by sustaining livelihoods based on artisanal fishing and tourism.

An overarching problem facing the Shimoni-Vanga Seascape is environmental degradation due to weak community organisational capacity in collective decision-making and action in building and maintaining the resilience of this socio-ecological production landscape and seascape (SEPLS). Previous institutional support to

T. Tschentscher (✉)
UNDP Consultant, Berlin, Germany
e-mail: tamara.tschentscher@undp.org

N. Chege
UNDP/GEF SGP Kenya, Nairobi, Kenya

H. Remaury
UNDP/GEF SGP, Paris, France

© The Author(s) 2023
M. Nishi, S. M. Subramanian (eds.), *Ecosystem Restoration through Managing Socio-Ecological Production Landscapes and Seascapes (SEPLS)*, Satoyama Initiative Thematic Review, https://doi.org/10.1007/978-981-99-1292-6_10

counteract biodiversity loss was significantly weak,[1] and where policies are appropriately targeted, e.g. in Community Managed Areas (CMAs), there is still low enforcement, as well as inadequate financial support and technical assistance. Although appropriate legal frameworks are in place, destructive methods of extraction and overexploitation of natural resources persist, particularly with regard to forests, mangroves, and fish populations. Destructive fishing and overexploitation of some fish species critical to ecosystem functioning, as well as unsustainable tourism practices, have particularly affected coral reefs and seagrass beds. Mangrove forests have faced additional threats from illegal logging and climate change.

Funded by the Global Environment Facility (GEF) and implemented by the United Nations Development Programme (UNDP), the community-based seascape management approach employed by the GEF Small Grants Programme (SGP) in Kenya aimed to restore ecosystem functions that form the basis of a mosaic of resource uses and enhance resilience of this socio-ecological production seascape. Three ecologically sensitive areas of global and national significance were selected for the implementation of this Sixth Operational Phase of the GEF SGP, one of which is the biodiversity-rich marine ecosystem of southern Kenya—the Shimoni-Vanga Seascape.

Through a participatory baseline assessment conducted in 2018, key stakeholders such as community members and leaders, civil society, academia, and national and county government agencies were engaged from the onset. They jointly identified key issues affecting the ecosystem and resource degradation, assessed capacities and needs, as well as determined priorities for community conservation actions.

Through partnerships with a multitude of stakeholders that ensured different interests were addressed and common conservation goals were pursued, the supported initiatives jointly contributed to revitalising this critical socio-ecological production seascape. Between 2018 and 2022, activities promoted mangrove forest and coral reef rehabilitation, eco-tourism enterprise development, sustainable fisheries and fish processing value chain development, and improved waste management systems. They have further contributed to strengthening multi-stakeholder collaboration, the capacities of Beach Management Units (BMUs), and improved monitoring, control, and surveillance of Locally Managed Marine Areas (LMMAs). Coral reef regeneration initiatives have piloted innovative restoration methods to support the healthy functioning of reef ecosystems, studying their strengths and weaknesses to determine the most community-accessible and effective models for scaling up and replication. These community-based initiatives are vital to creating a "society in harmony with nature" and conserving the rich local marine and terrestrial biodiversity (GEF SGP Kenya et al. 2018). This case study showcases these activities and documents knowledge and best practices that build the resilience of SEPLS by

---

[1] Poaching is driven by poverty and the accessibility of lucrative illegal markets. While Kenya is a signatory to the Convention on International Trade in Endangered Species of Wild Fauna and Flora (CITES) that bans trade in protected species, inadequate collaboration by law enforcement institutions poses a substantial challenge in enforcement of anti-poaching laws (NEMA 2017).

developing and diversifying livelihoods while enhancing biodiversity conservation and ecosystem services.

10.2   The Seascape

10.2.1   Geography

Located in Kwale County in the southern coastal region of Kenya, the Shimoni-Vanga Seascape covers an area of 860 km^2 (see Fig. 10.1; Table 10.1). Forming part of the transboundary Msambweni-Tanga seascape, it shares its northern Shimoni boundary with Msambweni (Funzi Bay) and its southern Vanga boundary with Tanzania. Seven villages are located along the seascape: Shimoni, Majoreni, Vanga, Jimbo, Kibuyuni, Wasini, and Mkwiro. The people of these villages largely depend on the seascape for their livelihoods. The latter two villages are located on Wasini Island, just off the coast from Shimoni village. Apart from Wasini Island, the seascape comprises the uninhabited Mpunguti ya juu, Mpunguti ya Chini, Mako Kokwe, and Sii coral islets (GEF SGP Kenya et al. 2018).

10.2.2   Biological Resources and Land Use

The seascape is rich in floral and faunal biodiversity, particularly corals, mangroves, and seagrasses, as well as a variety of marine mammals, crustaceans (shrimps, lobsters, and crabs), pelagic and demersal fishes, cephalopods (octopus, squids, and cuttlefish) and echinoderms (e.g. sea cucumbers), birds, and reptiles. Sandy beaches and estuaries are also found in the area with one permanent river (Ramisi) draining into the ocean. The importance of the seascape is underscored by a Marine Protected Area (MPA), situated within the Kisite-Mpunguti Marine National Park and Reserve, that demonstrates the highest coral diversity of Kenya's MPAs, with 203 species and a 50% coral cover (Obura 2012).

Such seascape ecosystems are highly interconnected, with coral reefs, mangrove forests, and seagrass beds depending on each other's health to maintain their sensitive ecosystem balance. Local people depend on this seascape for fishing, tourism activities, mangrove harvesting, and seaweed farming. Arable land with moderate rainfall and warm weather form ideal conditions for agricultural production in the area. Livestock rearing is also practiced in the hinterland of the area.

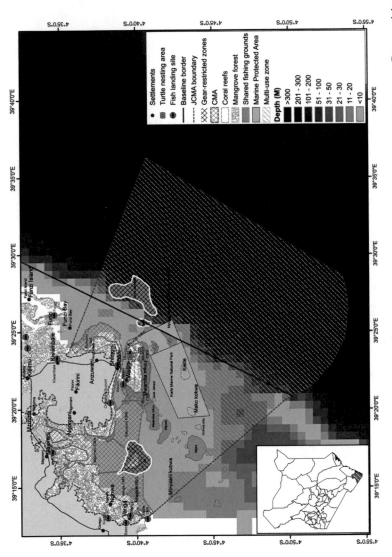

Fig. 10.1 The Shimoni-Vanga Seascape [Source: The Coastal and Marine Resource Development (COMRED) and Naturecom Group and the Regional Center for Mapping and Resource Development (RCMRD)]

Table 10.1 Basic information of the study site

Features of the study area	Description
Country	Kenya
Province/State	Kwale County
District	Lunga Lunga sub-county
Municipality	Vanga Ward
Area of analysis	Shimoni-Vanga Seascape
Size of the geographical area of analysis (hectare)	86,000
Size of the case study/project area (hectare)	Approx. 10,000
Dominant ethnicity(ies) of the project area, if appropriate	Wavumba
Number of direct beneficiaries (people)	2682
Number of indirect beneficiaries (people)	10,728
Geographic coordinates (latitude, longitude)	4°38′51.2″ S, 39°22.905′ E

Source: Authors, based on this study

10.2.3 Socio-Economic Context

The majority of the 18,000 people living within the Shimoni-Vanga Seascape depend mainly on fishing, farming, tourism, and trade for their livelihoods (GEF SGP Kenya et al. 2018). Some villages are home to marginalised and indigenous groups of Wakifundi and Wachwaka, as well as Makonde migrants from Mozambique.

Fishing is the predominant livelihood activity in all the villages, with nearly 100% of the population in Mkwiro village dependent on fishing, followed by fish trading (25%) (KCDP 2016a). Cash crop farming accounts for 22% of household incomes, while livestock production contributes 18% (GEF SGP Kenya et al. 2018). There have also been attempts to develop small-scale fish processing, particularly in Kibuyuni, Majoreni, and Shimoni. Tourism services such as boat tours and mangrove boardwalks help diversify livelihoods in the area.

About 2632 fishers are active in the Shimoni-Vanga Seascape. Migrant fishers, who are mostly found in Vanga and Jimbo, make up about 15% of local fishers (GEF SGP Kenya et al. 2018). Multiple fish species can be sourced with various types of gear, but illegal fishing gear such as spear guns and beach seines are frequently spotted. Fisherwomen only make up about 3% of total fishers and are mainly found in Kibuyuni, Mkwiro, and Wasini (see Fig. 10.2 for a map of Wasini Island). However, women are particularly active further along the fish value chain, particularly in processing and selling fish at the village level.

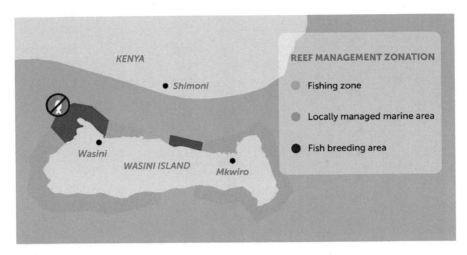

Fig. 10.2 Map of Wasini Island (Source: Prepared by Vrijlansier Guido Paap, REEFolution in 2017)

10.3   Challenges and Opportunities

10.3.1   Environmental and Social Challenges

To determine a comprehensive baseline and jointly set intervention priorities for SGP-supported, community-led initiatives with seascape stakeholders, existing key challenges and opportunities were assessed through extensive community consultations and a baseline assessment conducted in 2018. This process is explained in further detail in the next section. Table 10.2 summarises specific threats to target ecosystems and communities in the Shimoni-Vanga Seascape, as determined during these discussions (GEF SGP Kenya et al. 2018).

10.3.2   Opportunities for Ecosystem Conservation and Livelihood Development

One approach to improve the effectiveness of fisheries management in Kenya is the direct supervision of joint fisheries co-management areas (CMAs) and related management plans by local communities, notably Beach Management Units (BMUs) and county authorities (MALF 2017).[2] This type of arrangement is rooted in the legal framework of the Fisheries Management and Development Act No. 35 of

---

[2]The Government of the Republic of Kenya is composed of 47 counties, and each county has its own semiautonomous government.

Table 10.2 Specific threats to target ecosystems and communities in the Shimoni-Vanga Seascape

Mangroves	• Key threats include overexploitation of wood and non-wood products; conversion of mangrove areas to other land uses such as rice farming, infrastructure development, pollution effects, and sedimentation. • Climate change is also affecting the remaining mangroves to some extent, particularly through sea level rise, aridity, and influx of freshwater causing floods and sedimentation. • One of the key drivers of mangrove deterioration has been a lack of cohesion amongst stakeholders in the use and management of the resource. This has been exacerbated by a lack of recognition of customary rules and indigenous knowledge in governing subsistence use of mangrove resources.
Coral reefs	• As in many parts of the world, a major driver of coral decline is the rise of sea surface temperature that induces coral bleaching and mortality. • Over-harvesting of reef resources, particularly fish that keep the coral ecosystem balanced, has also impacted the health of local reefs. There are no controls on types of fish harvested. Destructive fishing practices such as seine nets and trampling of corals by speargun fishers causes physical damage to corals. • Tourism activities also frequently physically damage the corals that provide shelter to small juvenile and coral-based (cryptic) fish.
Beaches	• Although the seascape comprises some of the most renowned beaches in the world that are important not only for recreation, but also serve as nesting grounds for turtles, these beaches face many threats, such as trampling, infrastructure development, beach erosion, flooding due to sea level rise, shoreline changes due to rough seas, and discarding of litter. These threats mainly affect sea turtle nesting, as well as the recreational and aesthetic value of beaches.
Seagrass beds	• Key threats include destructive fishing (i.e. beach seines); overexploitation of triggerfish that feed on sea urchin predators that are a main threat to seagrasses; pollution; dredging and boating activities that uproot seagrasses; as well as climate change effects. To stem further degradation of seagrasses, the Kenya government developed a management and conservation strategy for coral reefs and seagrass ecosystems for 2014–2018.
Key conservation species	• Marine mammals (humpback whales, dolphins, and dugong), birds, whale sharks, and sea turtles—which form important tourism attractions—face numerous threats such as injury from boat strikes, especially during peak fishing and tourism seasons when there is increased boat traffic, incidental catches, and poaching (turtles and dugong).
Human well-being	• Key threats, especially for communities that largely depend on seascape resources, include lack of alternative livelihoods, loss of customary access rights, diminishing resource-dependent incomes (e.g. through reduced fish populations), competition between small-scale resource users and commercial users, low literacy levels, youth unemployment, loss of local and indigenous knowledge, lack of market access, and limited formal access to loans.

Source: Summarised from GEF SGP Kenya et al. (2018)

Fig. 10.3 Fishing boats at low tide (Photo: Dishon Murage, Seacology Foundation)

2016 as well as the 2007 Fisheries BMU Regulations. For long-term sustainability, this co-management system adopts an ecosystem approach to fisheries (EAF).

Since Kenya adopted an EAF framework in 2011, several area-, species-, and gear-based management plans have been developed using this approach (Garcia et al. 2003).[3] These include the Lobster Fishery Management Plan, Small-Scale Purse Seine Fishery Management Plan, and Marine Aquarium Fisheries Management Plan.

In Shimoni-Vanga, seven BMUs, each representing one seascape community, jointly co-manage the CMA seascape to promote ecosystem conservation, optimise social and economic benefits for local communities, and improve governance and management of the seascape. This includes ensuring compliance with local and national resource use regulations, as well as approval of fishing vessel registrations and fishing licenses (Fig. 10.3 shows a local fishing boat).

The 5-year Joint Co-management Area Plan (JCMAP) (2017–2021) included seven village-based Co-management Area Plans (CMAPs) that complemented the Kisite-Mpunguti Marine Protected Area Management Plan (2015–2025) (GEF SGP Kenya et al. 2018; KCDP 2016b). However, the managing BMUs struggled with insufficient capacity to effectively conduct Monitoring, Control, and Surveillance (MCS), which, coupled with poor inter-agency collaboration and governance, has

---

[3]The EAF was adopted by the FAO Committee on Fisheries (COFI) in response to the need to implement sustainable development principles, the Convention on Biological Diversity, and the Code of Conduct for Responsible Fisheries (FAO 1995).

resulted in weak enforcement. Despite these weaknesses, this community-management framework represents a promising foundation to further build the socio-economic resilience of this production seascape with global significance (COMRED and Naturecom Group 2020b).

10.4 Activities, Achievements, and Impacts

Collective action by the various stakeholders to strengthen socio-ecological resilience for the sustainable use and management of coastal marine resources is key to broad acceptance of the outcomes of collective action and achieving sustainable benefits for humans and ecosystems (COMRED and Naturecom Group 2021). Emphasis on support from various actors to community organisations for development and implementation of adaptive management projects reinforces the positive results of the collaborative efforts. Based on this understanding and building on the joint CMA approach to resource management, the GEF Small Grants Programme (SGP) in Kenya employed a highly participatory, community-based seascape management approach as its core programming framework for its 6th Operational Phase (OP6).

To ensure various actors and stakeholders were closely engaged in the design, implementation, documentation, and monitoring of activities, GEF SGP Kenya worked with a broad spectrum of stakeholders across government, civil society, the private sector, and academia.

10.4.1 The Seascape Approach

As identified during consultations with key stakeholders within the baseline assessment conducted in 2018 by the GEF SGP, an overarching challenge in achieving sustainable use and management of coastal marine resources and other related natural resource systems has been the lack of a coordinated approach to harness the connectivity of the ecosystems. A segregated approach can result in an intervention in which one component is beneficial to certain ecosystem segments but detrimental to others, which can cause conflicts of interest for different groups (COMRED and Naturecom Group 2021).

The seascape approach in Shimoni-Vanga aimed to address a mosaic of seascape resource uses in a holistic manner. Management of such a mosaic seascape with resources that are highly interconnected required an integrated approach that facilitated and promoted interaction among resource users. The seascape approach encouraged cross-community interactions and synergies among community projects, enabling harmonisation of activities to optimise protection, manage trade-offs, and ensure sustainability.

10.4.2   Stakeholder Engagement and Use of Resilience Indicators

Local communities have a great interest in ensuring the long-term resilience of their SEPLS in general, particularly when the majority of the local population depends on ecosystem services from landscape or seascape for their livelihoods. They also often hold the key to unlocking the most suitable solutions by coupling traditional knowledge with modern best practices when given the chance to participate.

Starting with a participatory baseline assessment applying the Indicators of Resilience in SEPLS (Bergamini et al. 2014), community members involved in environmental conservation in the seascape were directly engaged from the onset. Community members and leaders from seven villages within the Shimoni–Vanga seascape, along with four community-based organisations representing both men and women in each village, and representatives from government agencies, the private sector, and local NGOs participated in the baseline assessment in 2018. This process ensured joint identification of: resources within the seascape, their socio-ecological resilience, and key issues affecting their management, as well as community conservation priorities and proposed actions.

The Coastal and Marine Resource Development (COMRED) and Naturecom Group was identified as a strategic partner to oversee the baseline assessment, as well as to engage in development and implementation of the seascape strategy. Since its establishment in 2006, this non-profit organisation has promoted practical solutions to problems faced by coastal and marine environments and communities in achieving sustainable development. COMRED was also one of the key consultants engaged by the national government to develop the Joint Co-Management Area Plan for the Shimoni-Vanga seascape.

10.4.3   Seascape Strategy and Community-Led Landscape Projects

The baseline assessment and community consultations gave rise to the Shimoni-Vanga Seascape Strategy, which set out five landscape outcomes and associated indicators to measure progress towards these outcomes (see Table 10.3 for an overview of Seascape Outcomes, Key Performance Indicators and suggested activities jointly determined by stakeholders and outlined in the Shimoni-Vanga Seascape Strategy). The strategy formed a key element of the seascape planning process, where seascape communities generated a shared vision of what a more resilient local seascape would look like and determined what actions would be required to realise this vision (UNDP 2016).

Table 10.3 Overview of seascape outcomes, key performance indicators, and suggested activities

Seascape outcomes	Key performance indicators	Suggested activities
1. Integrity of habitats and biodiversity within the seascape is enhanced	• Number of hectares restored, protected, and managed • Species cover percentage, relative abundance, and distribution • Number of new tree stumps in the seascape	• Rehabilitate degraded habitats (e.g. seagrass beds, mangroves, coral reefs) • Create awareness on the value of key habitats, their connectedness, threats, and consequences of their degradation • Strengthen management and resource monitoring of existing Co-management Areas (CMAs) and Joint Co-management Areas (J-CMAs) • Establish temporal closures to allow resource rejuvenation and biomass build-up
2. Pressure on mangrove resources is reduced	• Number of alternative energy apparatuses installed in the seascape • Number of energy saving and efficient apparatus put in use in the seascape	• Introduce the use of alternative energy sources • Reduce dependence on natural resources by using energy-saving technologies
3. Livelihoods of communities living around the seascape are strengthened and diversified	• Number of successful income-generating activities, both new and strengthened initiatives • Number of jobs created	• Introduce and enhance innovative income-generating activities as an incentive to conservation • Value chain development initiatives in fishing, tourism, agriculture, aquaculture, and forestry • Assist communities to access climate change opportunities, such as sale of carbon credits and climate change funding • Link community-based small-scale enterprises to financing opportunities and civil society organisation (CSO)-private sector partnerships • Introduce renewable energy and energy efficiency solutions to cut costs and dependence on natural resources
4. Knowledge management among different actors is enhanced and shared	• Number of best practices and lessons learnt shared and adopted among seascape stakeholders • Number of documentation	• Promote actions that preserve local knowledge (ecological, cultural, religious) • Create community awareness that incorporates local

(continued)

Table 10.3 (continued)

Seascape outcomes	Key performance indicators	Suggested activities
	tools/materials developed and shared amongst actors and wider stakeholders	knowledge on conservation, livelihood options, climate change, democratic and accountable governance • Establish information centres and repositories that incorporate local knowledge • Support collection, dissemination, and sharing of lessons learnt and best practices • Capture and document indigenous knowledge (including marginalised and vulnerable groups) on use and management of seascape resources
5. Institutional seascape governance capacity is strengthened	• Number of community groups effectively participating in seascape decision-making • Proportion of community representation in seascape decision-making	• Strengthen Monitoring, Control, and Surveillance (MCS) of existing CMAs and J-CMAs • Establish and support a framework for a multi-stakeholder platform to enhance coordination and cooperation in project activities • Improve functioning of BMUs through capacity building on governance • Enhance equity and gender considerations where women and youth, and vulnerable and marginalised groups are involved in decision-making • Strengthen recognition and role of customary access rights to resources • Facilitate actions that promote building of social capital to achieve cooperation

GEF SGP Kenya et al. (2018)

10.4.4 Achievements and Impacts to Date

Based on this seascape strategy and the roadmap for the selection of project proposals, 16 community-led initiatives were competitively selected and received financial and technical support from GEF SGP Kenya to contribute towards improved resource management, livelihood development, and ecosystem restoration (COMRED and Naturecom Group 2020b). Activities implemented based on an integrated approach included: capacity building for improved Monitoring, Control, and Surveillance (MCS), particularly of the Locally Managed Marine Areas (LMMAs)[4]; expansion and demarcation of LMMAs; mangrove planting and coral reef rehabilitation; eco-tourism enterprise improvements; fish value addition and post-harvest handling; waste management and value addition; and education and awareness through art and performance (UNDP 2021).

Key achievements are highlighted as follows:

Environmental Awareness Harnessing the Kenyan enthusiasm for music and arts, the Mchongo youth group has been successfully using dance, theatre, poetry, and songs to entertain, but essentially to educate and create awareness among local communities on a variety of topics. Through SGP support, this group was able to acquire technical equipment to improve their communication products. Gaining wide recognition among communities was instrumental for the group in communicating messages on conservation through art, networking, and YouTube videos aimed at raising awareness on the importance of personal responsibility for care of the environment (UNDP 2021).

Coral Reef Rehabilitation The Wasini and Mkwiro BMUs piloted innovative coral reef restoration projects that are accelerating the slow, natural recovery of reefs. With close engagement of local communities, projects worked to enhance reef structure through artificial reef constructions onto which coral fragments grown in nurseries were transplanted (see Figs. 10.4 and 10.5). In Mkwiro, a project collected 30 species of common branching coral genera, which were multiplied over 2 years between August 2019 and August 2021 until the nursery contained 10,000 fragments. Since then, fragments can be harvested on an annual basis. Between 2019 and 2021, about 3000 coral fragments were placed on 750 artificial reef structures each year. In Wasini, 8300 coral fragments were raised in mid-water table nurseries for 6–8 months following a coral gardening concept before being transplanted on denuded reef substrates or on artificial reef structures (see Fig. 10.6). Materials used for nursery fixtures and artificial reef modules were locally sourced. These coral restoration sites have brought significant benefits to local communities. Ecological benefits include healthy reefs that have become fish aggregation and breeding zones, leading to recovery of fish populations. Activities further generated employment opportunities, including those for 35 community members engaging in coral restoration activities and 11 trained and employed reef rangers. Fish spillover from aggregation zones has benefitted local subsistence fishers. Apart from increased fisheries income, local tourism is benefitting as the healthy reef sites and restored fish populations attract divers and snorkelers. There have already been turtle sightings that had not occurred in years. Finally, these initiatives have considerably contributed to increased awareness and community conservation stewardship,

---

[4] An LMMA is an area of nearshore waters and coastal resources that is largely or wholly managed at a local level by the coastal communities, land-owning groups, partner organisations, and/or collaborative government representatives who reside or are based in the immediate area.

Fig. 10.4 Nursery of moulds being deployed at sea (Photo: Dishon Murage, Seacology Foundation)

generating vital research data and lessons for successful coral restoration projects and enhancing community ownership and partnerships among key stakeholders (REEFolution Foundation et al. 2021).

Mangrove Restoration Kenya Forest Service (KFS), a government agency, took a leading role in advising local communities on mangrove nursery establishment and mangrove planting. Over 6300 mangrove seedlings were planted by the Majoreni Beach Management Unit (BMU), Jimbo BMU, and the Wasini Women Group. These have contributed considerably to improving fisheries production as well as carbon capture and coastal protection (see Fig. 10.7). Regular monitoring of replanted mangroves was undertaken in areas where there is a functional Participatory Forest Management Plan (PFMP) and a Management Agreement (MA) developed between the community and KFS. The PFMPs and MAs require the community and partners to identify areas in need of replanting, undertake replanting, and conduct regular monitoring of replanted areas. So far, replanted mangroves have shown a 50% survival rate. These activities have created 110 casual jobs for local communities in mangrove nursery establishment, operation, and on mangrove seedling plantations for mangrove forest restoration (UNDP 2021).

Fish Value Chain Addition Many women in coastal Kenya make a living as fish merchants and have done so for a long time. However, incomes have remained low. To improve the capacities and livelihoods of these still marginalised women, the

Fig. 10.5 Coral fragment nursery (Photo: Dishon Murage, Seacology Foundation)

Fig. 10.6 Overview of coral gardening structures used in Mkwiro LMMA: Coral nursery tree (**a**), Bottle reef unit (**b**), Layered cake (**c**), Metal cage (**d**) (Photos: Ewout Knoester, REEFolution Foundation)

community-based organisation Indian Ocean Water Body (IOWB) led activities with four BMUs in Mkwiro, Shimoni, Vanga, and Jimbo. IOWB provided capacity-building training on fishing quality control (handling), value addition, marketing, sales, and entrepreneurship development to 112 women. It also supported women in the purchase of equipment such as cooling boxes, freezers, and weighing scales to further improve their distribution and sales and helped upgrade Pwani Fish Marketing's fish processing facility. With the resulting increased incomes and enhanced financial independence through fish trade, the women have noted a higher level of pride and satisfaction (UNDP 2021).

Fig. 10.7 Mangrove rehabilitation (Photo: Jimbo Beach Management Unit)

Waste Management The Center for Justice and Development (CEJAD)—a national non-governmental organisation (NGO) and GEF SGP grantee partner—partnered with the communities of Mwkiro and Wasini through the Mkwiro Eco-friendly Conservation Group and Wasini Women Group to raise awareness on the threat of marine plastic pollution, organise beach clean-ups, and promote locally suitable waste recycling mechanisms that support sustainable livelihoods. They purchased solid waste bins and established a demonstration centre for recycling solutions. Following the philosophy "*taka ni mali*" ("waste is money"), CEJAD facilitated the training of 40 community members, 30 of them women, and provided equipment and tools to turn plastic waste into sellable artefacts such as key holders, mats and hats, and bracelets (see Figs. 10.8 and 10.9). These have been exhibited and offered for sale at the demonstration centre. Through installation of a waste baling machine and a partnership with a private sector company to recycle plastic, further employment opportunities have been created for waste collection, sorting, cleaning, baling, transportation to recyclers, and facility security. The baler has been a game changer as it tremendously cuts costs for waste processing and transportation. Increased awareness on pollution impacts coupled with employment and income generation opportunities have led to broad community support for these initiatives. Local communities have celebrated World Oceans Day, conducting beach clean-ups with grantee partners. To further improve waste management governance in the long term, the Wasini Women Group established a community

Fig. 10.8 Hats and mats made by the Mkwiro Eco-friendly Conservation Group (Photo: Dishon Murage, Seacology Foundation)

Fig. 10.9 Artefacts made from recycled marine waste by the Mkwiro Eco-friendly Conservation Group (Photo: Dishon Murage, Seacology Foundation)

waste management committee that developed rules on waste management to reduce pollution, and CEJAD contributed to a solid waste management policy for the county government of Kwale (UNDP 2021).

Eco-tourism Development Shimoni has a rich, historical past, albeit a sad one. At the height of the slave trade in the 1750s, it was a bustling town as a slave-holding port for East Africa's slave trade. The holding pens were the naturally existing caves, which today are tourist attractions. The local community in partnership with the National Museums of Kenya enhanced tourism infrastructure by installing solar lights in the caves, establishing a nature trail in the nearby Shimoni forest, and erecting a viewing platform next to the caves. With the newly repaired road, and the impending construction of a fishing port in Shimoni, tourism traffic is bound to grow. Additionally, Wasini and Mkwiro community members were trained as dive guides to offer tours to give visitors chances to see the growing number of fish species resulting from the successful coral rehabilitation activities, as well as improved management of the locally managed marine area. Through a partnership between the local boat owners association and Kenya Wildlife Service, community members can now offer boat tours to Kisite-Mpunguti Marine National Park, which is known for its dolphin sightings.

CMA Management Capacities Under the leadership of the Kenya Fisheries Service and the Fisheries Department of Kwale County Government, NGO partners conducted trainings in MCS tools and processes to strengthen enforcement capacities of BMUs. In August and September 2020, 76 executive and patrol sub-committee members of the BMUs (including 24 women) received theoretical and practical training in pre-patrol, patrol, and post-patrol procedures, fish catch data collection and monitoring, fish handling and quality assurance, documentation and reporting practices, and patrol planning logistics. The trainings were led by officers from Kenya Fisheries Service, Kwale County Fisheries Department, Kenya Coast Guard Service, Kenya Wildlife Service and Kenya Forest Service, with support from COMRED as a lead strategic partner. As a result of these trainings, BMUs reported an increased number of licensed fishers and registered fishing crafts—for example, 190 fishers licensed in Vanga in 2020 compared to 166 in 2019, and 60 registered fishing boats in 2020 compared to 41 in 2019 (see Fig. 10.10 for a local canoe in operation). The Kibuyuni BMU reported 105 fisher licenses in 2020, a significant increase from 29 in 2019 (COMRED and Naturecom Group 2020b). Additionally, enforcement of BMU by-laws has been improved. By-laws ban fishing in certain zones during spawning seasons and stipulate fishing net mesh sizes to ensure that only mature fish are caught, for example. This is to some extent a return to more traditional practices.

Participatory Governance One of GEF SGP Kenya's major targets during this operational phase was to establish a multi-stakeholder platform to foster long-term working relationships and create a shared vision among seascape stakeholders with respect to coastal and marine resource use, management, and development. This multi-stakeholder forum for the Shimoni-Vanga Seascape was established with

Fig. 10.10 Local canoe used for transport in Majoreni fish landing station (Photo: Dishon Murage, Seacology Foundation)

support from COMRED and Naturecom Group as strategic partners, as well as the Government of Kenya and County Government of Kwale (COMRED and Naturecom Group 2020a). It was officially inaugurated in January 2020, co-chaired by the Ministry of Interior and Coordination of the national government (Office of Deputy County Commissioner, Lunga Lunga sub-county) and the County Government of Kwale (Sub-county administrator, Lunga Lunga sub-county). Membership in the multi-stakeholder platform is based on the area of residence and interest in the seascape. Members include: (i) statutory agencies such as Kenya Wildlife Service, Kenya Fisheries Service, National Environment Management Authority (NEMA), and statutory community-level agencies, such as Beach Management Units (BMUs) and Community Forest Associations (CFAs); (ii) non-state actors—NGOs, CBOs, and the private sector, and (iii) observers—parties who may have an interest in the seascape (COMRED and Naturecom Group 2021).

10.5 Discussion

10.5.1 Key Innovations for Replication

The active coral reef restoration projects led by the Wasini and Mkwiro BMUs, piloting innovative methods and techniques, have generated remarkable results and impacts. Partnering with the REEFolution Foundation as well as Wageningen University and Research, the Coastal Oceans Research and Development—Indian Ocean (CORDIO) East Africa, and the Kenya Marine Fisheries and Research Institute (KMFRI), the BMUs and COMRED piloted and evaluated different coral fragment nursery types, artificial reef structures, and gardening techniques as mentioned earlier. Although these methods have been tested before in other contexts,[5] the strengths and weaknesses of different techniques had not yet been compared and documented. At both project sites, SGP-supported projects tested different coral species for nurseries and transplantation, monitoring their growth and survival rates with different stressors (such as bleaching events) over time and artificial reef construction types (see Fig. 10.11 for examples of coral fragments' growth). This data was recorded to identify the most resilient species and most suitable growing and transplantation methods. One of the key conclusions by the Mkwiro BMU highlighted that the coral-tree design is the most cost-effective nursery technique, where a full-grown coral fragment can be reared for about US\$0.33 per fragment, including costs of nursery construction, diving costs, maintenance, and labour.[6] Monitoring will continue with assessments of long-term and ecologically significant changes in coral cover and fish populations on a biannual basis. The invaluable knowledge and lessons generated from these initiatives will aid decisions on the most feasible and successful coral reef restoration techniques for future community-based initiatives (REEFolution Foundation et al. 2021).

10.5.2   Progress at the Seascape Level

The establishment of the multi-stakeholder platform has been an important milestone in ensuring engagement of communities and the various local stakeholders of the coastal and marine resources in any related decision-making processes. It enhanced effective and collective decision-making and strengthened the conservation and management of natural resources by aligning and coordinating activities. This platform also helped by informing stakeholders on implementation of government plans and initiatives and ensuring that the voices and needs of different interest groups informed joint decisions for shared benefits. By incorporating and aligning new initiatives and strategies to existing government plans, coordination, and cross-sectoral synergies can be enhanced (COMRED and Naturecom Group 2020a). Committees for different thematic areas are in the process of being established as

---

[5] Coral reef restoration is a relatively new approach to reef restoration both in Kenya and in the region. However, currently, a number of initiatives in reef restoration using similar approaches are being undertaken in other coastal areas in Kenya, such as Kuruwitu (Kilifi), Pate (Lamu), and Diani (Kwale). One of the most successful reef restoration initiatives is on Cousin Island, Seychelles, while other reef restorations are currently being undertaken in Tanzania and Mozambique.

[6] The costs for the other artificial reef structures were as follows: \$15.21 per Bottle Reef Unit (BRU), \$43.78 per Cake and \$59.12 per Cage (including welding costs).

Fig. 10.11 Example of coral performance planted on artificial reef units for commonly used coral species in the Mkwiro project. The left column shows pictures just after planting in 2019 and the right column of pictures shows the same reef units in 2021 (Photos: Ewout Knoester, REEFolution Foundation)

part of the forum, including those on research, security and surveillance, livelihoods and community, management, fundraising, and cross-cutting areas such as gender and health.

On 9 December 2021, the Kisite-Mpunguti Marine National Park and Reserve, located within the Shimoni-Vanga Seascape, made history as Kenya's first Blue Park. It was awarded the gold-level Blue Park Award by the Marine Conservation Institute for achieving the highest standards for marine life protection and management. The Blue Park Award recognises outstanding efforts by nations, MPA managers, and local community members to effectively protect marine ecosystems now and into the future. There are only 21 Blue Parks around the world, designed to protect and regenerate ocean biodiversity.

10.5.3   Lessons Learned and Moving Forward

The MCS trainings also involved discussion of challenges and lessons learned, as well as identification of needs and a way forward. Analyses and discussions highlighted that Illegal, Unreported, and Unregulated (IUU) fishing has predominated Shimoni-Vanga's small fisheries industry. Reasons include lack of awareness and ignorance of fisherfolks regarding laws and regulations. There is also insufficient understanding that unsustainable practices will affect the very resources that the fisherfolk harvest and depend on. Initiatives will need to build on early efforts to sensitise local communities concerning impacts on their livelihoods and support income generation by value chain addition activities and other alternative livelihood opportunities.

Key lessons and recommendations include the need for regular follow-up and support in implementation and monitoring of progress and translation of training modules to Swahili, which is better understood by most BMU members. Discussions also highlighted that support from and engagement of different stakeholders, including the national and county government as well as civil society actors, is essential to further improve BMU operations (COMRED and Naturecom Group 2020b).

Moving forward, the possibility of joint patrols among all the Shimoni-Vanga BMUs and the engagement of enforcement agencies, e.g. Fisheries and Kenya Coast Guard, to improve the effectiveness of this mechanism is being explored among stakeholders. Also, integrity trainings and protocols are being elaborated to avoid leakage of patrol information. Furthermore, initiatives will promote clear demarcations of designated fishing areas.

To further reduce illegal fishing practices, county government support to communities for acquisition of sufficient adequate fishing gear will be needed. Fish merchants also need to be further sensitised on avoidance of purchasing undersized fish and fish species that are not for human consumption. BMUs have already discussed and stand ready to support the identification and development of locally suitable income-generating activities for improved local livelihoods and further support of MCS activities in the seascape (COMRED and Naturecom Group 2020b).

10.6 Conclusions

Over a period of three and a half years from mid-2018 to early 2022, the Small Grants Programme, implemented by UNDP, provided technical assistance and financial support from the GEF to 16 local organisations to build the social, economic, and ecological resilience of the Shimoni-Vanga Seascape. An integrated, community-based seascape approach was taken to engage local communities and other key stakeholders in the seascape in a highly participatory seascape planning process. This allowed for the development and implementation of adaptive management projects in ways that ensured that benefits, priorities, and activities, for both the global environmental agenda and local sustainable development, were effectively aligned with the different interests of seascape resource users. The wide range of activities were designed to complement each other and generate synergistic results at the seascape level, promoting marine and coastal biodiversity conservation, as well as building organisational and governance capacity for enhanced financial and institutional sustainability. Despite this short implementation period, promising results have already been recorded. Although there is still much that can and needs to be done, the implementation of this integrated seascape approach that emphasises community priorities have demonstrated that ecosystem restoration and livelihoods can be effectively revitalised.

References

Bergamini N, Dunbar W, Eyzaguirre P, Ichikawa K, Matsumoto I, Mijatovic D, Morimoto Y, Remple N, Salvemini D, Suzuki W, Vernooy R (2014) Toolkit for the indicators of resilience in socio-ecological production landscapes and seascapes. UNU-IAS, Biodiversity International, IGES, UNDP, Rome

Coastal and Marine Resource Development (COMRED) & Naturecom Group (2021) Innovations for a sustainable ocean: the Shimoni-Vanga multi-stakeholder's forum—promoting social impact and sustainable management of the Shimoni-Vanga Seascape. COMRED, Mombasa

Coastal and Marine Resource Development (COMRED), Naturecom Group (2020a) Memorandum for the establishment of the Shimoni-Vanga Seascape multi-stakeholders forum. COMRED, Mombasa

Coastal and Marine Resource Development (COMRED), Naturecom Group (2020b) MCS training report for the Shimoni-Vanga BMUs. COMRED, Mombasa

Food and Agriculture Organization of the United Nations (FAO) (1995) Code of conduct for responsible fisheries. FAO, Rome

Garcia SM, Zerbi A, Aliaume C, Do Chi T, Lasserre G (2003) The ecosystem approach to fisheries. Issues, terminology, principles, institutional foundations, implementation and outlook. Food and Agriculture Organization of the United Nations (FAO) Fisheries Technical Paper, No 443, Rome

GEF SGP Kenya, Coastal and Marine Resource Development (COMRED), Naturecom Group (2018) Seascape strategy for building social, economic, and ecological resilience—Shimoni-Vanga Seascape. GEF SGP Kenya, Nairobi

Kenya Coastal Development Project (KCDP) (2016a) Fisheries and socio-economic assessment of Shimoni-Vanga area. Baseline report. Kenya South Coast, KCDP

Kenya Coastal Development Project (KCDP) (2016b) Ecological risk assessment for the development of a Joint Co-Management Area (JCMA) plan in the Shimoni-Vanga Area, South Coast Kenya. State Department for Fisheries and Blue Economy report. KCDP, Kenya South Coast

Ministry of Agriculture, Livestock and Fisheries (MALF) (2017) The Shimoni-Vanga Joint Fisheries Co-Management Area Plan (JCMAP): a five-year co-management plan to guide the development of Shimoni-Vanga joint fisheries co-management area. MALF, Government of Kenya, Kwale County

National Environment Management Authority (NEMA) (2017) State of the coast report II: enhancing integrated management of coastal and marine resources in Kenya. NEMA, Government of Kenya, Nairobi

Obura D (2012) The diversity and biogeography of Western Indian Ocean reef-building corals. PLoS One 7(9):e45013

REEFolution Foundation, Wageningen University and Research (WUR), Coastal Oceans Research and Development—Indian Ocean (CORDIO) East Africa, Coastal and Marine Resource Development (COMRED), Kenya Marine Fisheries and Research Institute (KMFRI), Mkwiro Beach Management Unit (BMU), Wasini Beach Management Unit (BMU) (2021) Coral reef rehabilitation in Shimoni-Vanga Seascape: a case study of Mkwiro and Wasini LMMAs—developing community-accessible rehabilitation methods on degraded reefs. REEFolution Foundation, Oosterbeek

United Nations Development Programme (UNDP) (2016) A community-based approach to resilient and sustainable landscapes: lessons from phase II of the COMDEKS programme. UNDP, New York

United Nations Development Programme (UNDP) (2021) Project implementation report (PIR) 2021, sixth operational phase of the GEF SGP in Kenya. UNDP, New York

The opinions expressed in this chapter are those of the author(s) and do not necessarily reflect the views of UNU-IAS, its Board of Directors, or the countries they represent.

Open Access This chapter is licenced under the terms of the Creative Commons Attribution-NonCommercial-ShareAlike 3.0 IGO licence (http://creativecommons.org/licenses/by-nc-sa/3.0/igo/), which permits any noncommercial use, sharing, adaptation, distribution and reproduction in any medium or format, as long as you give appropriate credit to UNU-IAS, provide a link to the Creative Commons licence and indicate if changes were made. If you remix, transform, or build upon this book or a part thereof, you must distribute your contributions under the same licence as the original. The use of the UNU-IAS name and logo, shall be subject to a separate written licence agreement between UNU-IAS and the user and is not authorised as part of this CC BY-NC-SA 3.0 IGO licence. Note that the link provided above includes additional terms and conditions of the licence.

The images or other third party material in this chapter are included in the chapter's Creative Commons licence, unless indicated otherwise in a credit line to the material. If material is not included in the chapter's Creative Commons licence and your intended use is not permitted by statutory regulation or exceeds the permitted use, you will need to obtain permission directly from the copyright holder.

Chapter 11
Engaging Local People in Conserving the Socio-Ecological Production Landscape and Seascape by Practicing Collaborative Governance in Mao'ao Bay, Chinese Taipei

Jyun-Long Chen, Kang Hsu, Chun-Pei Liao, Yao-Jen Hsiao, and En-Yu Liu

11.1 Introduction

11.1.1 Background

Chinese Taipei's fishing industry has been traditionally based in harbours, and fishing settlements have gradually expanded into their surrounding areas. Accordingly, clusters of fishing communities became typical of Chinese Taipei's coastal regions. Coastal fishery is a major source of income and livelihoods for local communities along the coast. However, intensified activities (e.g. fisheries aimed

J.-L. Chen (✉)
Marine Fisheries Division, Fisheries Research Institute, Council of Agriculture, Keelung, Taiwan

General Education Center, National Taiwan Ocean University, Keelung, Taiwan
e-mail: jlchen@mail.tfrin.gov.tw

K. Hsu
Marine Fisheries Division, Fisheries Research Institute, Council of Agriculture, Keelung, Taiwan

Department of Bio-Industry Communication and Development, National Taiwan University, Taipei, Taiwan

C.-P. Liao
Marine Fisheries Division, Fisheries Research Institute, Council of Agriculture, Keelung, Taiwan

Y.-J. Hsiao
Institute of Applied Economics, National Taiwan Ocean University, Keelung, Taiwan

E.-Y. Liu
Institute of Oceanography, National Taiwan University, Taipei, Taiwan

© The Author(s) 2023
M. Nishi, S. M. Subramanian (eds.), *Ecosystem Restoration through Managing Socio-Ecological Production Landscapes and Seascapes (SEPLS)*, Satoyama Initiative Thematic Review, https://doi.org/10.1007/978-981-99-1292-6_11

at increasing catches, improved fishing efficiency, and development of more efficient fishing gear and fishing methods) (Worm et al. 2009) and climate change have caused a reduction in fishery resources and the destruction of marine ecosystems (Worm et al. 2009; Zhou et al. 2010), especially in coastal areas (Huang and Chuang 2010). Given that marine resources are decreasing, the Fisheries Agency of the Council of Agriculture, the national governing body in charge of Chinese Taipei's fishery management, began promoting the Coastal Blue Economy Growth (CBEG) programme (Lee 2015) in 2015 to improve the marine ecosystem and enhance fishery resources, adopting a marine stock enhancement approach. Mao'ao Bay was designated as Chinese Taipei's first sea farming demonstration zone with a community-based sea farming (CBSF) project to begin restoring ecosystems. The project comprised various measures for promoting sustainable fisheries, including pre-planning, stock enhancement, and community-based co-management.

The Mao'ao CBSF project offers important suggestions for building a collaborative governance framework, by involving partners across all levels of government, fishery associations, and residents (community), and for developing community-based marine resource restoration actions that may be feasible in other contexts (Chen et al. 2020). Useful and beneficial information based on the experiences and framework of promoting the CBSF project has continued to inspire decision makers and natural resources managers.

In recent years, the Forestry Bureau of the Council of Agriculture (FBCA) has been working with the public sector, universities, nonprofit organisations, and local communities to promote the ecological restoration of critical habitats and enhance ecosystem services in low-elevation mountainous ecosystems (e.g. rice terraces and wetlands) based on the concept of the Satoyama Initiative. In 2018, based on the need for maintaining ecosystem services derived from river basins and the integrity of connections among forests, rivers, human settlements, and seas in the natural and rural areas of Chinese Taipei, the FBCA launched a policy programme entitled, "Establishment of a National Ecological Conservation Green Network". The programme explores integrated approaches to the conservation, revitalisation, and sustainability of socio-ecological landscapes and seascapes (SEPLS). During the initial stage of implementation, expectations for coastal areas within SEPLS have grown, and the Fisheries Research Institute (FRI) has cooperated with the programme since 2019. The FRI started investigating coastal SEPLS, considering how to respond to current trends that centre on resource conservation management and local economic development. FRI researchers began to involve local people alongside government agencies, research institutes and academia, and other stakeholders, in the development of a set of measures for conserving coastal SEPLS.

11.1.2   Socio-Economic and Environmental Characteristics of the Area

Mao'ao Bay is located in the Gongliao District of New Taipei City, on the north-eastern coast of Chinese Taipei. It is home to outstanding fishing grounds that exhibit a rich diversity of marine species due to the three streams that flow into the bay, transporting essential nutrients (Fig. 11.1). Additionally, the shape of the bay

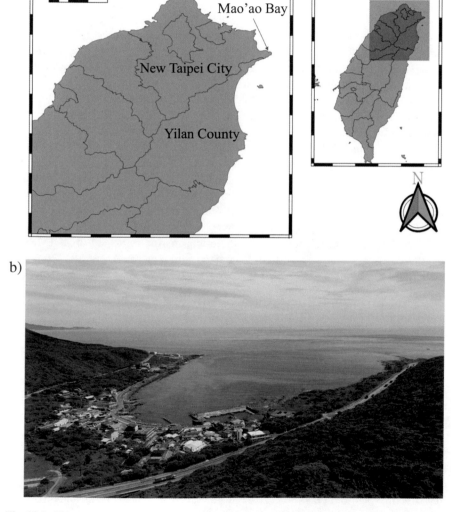

Fig. 11.1   The geographical location (**a**) and an aerial view (**b**) of the Mao'ao community [source: (**a**) Prepared by authors, (**b**) Photo Credit: Tso, Cheng-Wei]

Table 11.1 Basic information of the study area

Country	Chinese Taipei
Province	New Taipei City
District	Mao'ao Village, Gongliao District
Size of geographical area (hectare)	180
Dominant ethnicity	Han Chinese
Size of case study/project area (hectare)	73.8
Number of direct beneficiaries (people)	200
Number of indirect beneficiaries (people)	2000
Geographic coordinates (latitude, longitude)	25°00′44.7″N 121°59′29.3″E

Source: Prepared by authors

provides a great shelter for marine species to live, thereby fostering a stable ecosystem. Notably, small abalones (*Haliotis diversicolor*) and abalones (*Haliotis discus*) are cultured there, allowing the community to promote these products as delicacies and specialties. In addition, the local government has begun a basic rehabilitation of economically important fishery resources, including breeding and releasing pharaoh cuttlefish (*Sepia pharaonis*), small abalones (*H. diversicolor*), and abalones (*H. discus*).

Mao'ao is a small fishing village with approximately 200 residents, located close to the Mao'ao fishing harbour (within Mao'ao Bay) (Table 11.1). Fishery is a traditional practice in the Mao'ao Bay. The site exhibits smaller-scale fisheries, such as pole and line fishing, gillnetting, and harvesting. Additionally, the community relies on local produce, such as *Gelidium* spp., *Eucheuma serra*, and *Grateloupia filicina* seaweeds and processed fish roe. Several regulatory mechanisms for resource management exist, including marine protected areas (MPAs), fishing controls (fishing gear restrictions), size restrictions, and closures. Moreover, artificial reefs have been deployed outside the perimeter of Mao'ao Bay, further expanding the favourable conditions for rich marine life.

11.1.3 Objective and Rationale

It is essential to support both human well-being and the sustainable management of natural resources by understanding the dynamics of complex human–nature interactions within coastal social-ecological systems (Gain et al. 2019). Based on the definition of Satoumi, which is a "coastal sea preserved by humans for their survival with nature and culture" (Yanagi 2013, p. 113), the synergetic coexistence of humans and nature within coastal ecosystems is a central tenet of the Satoumi concept. Regarding the sustainable development goals (SDGs), the FRI emphasises SDG 14 (life below water), SDG 8 (economic growth), SDG 11 (sustainable cities and communities), and SDG 12 (responsible consumption and production).

Considering these goals, the environmental, social, and economic aspects of coastal communities should be integrated.

The FRI has accumulated experience over the long term via its research on and development of fisheries and has the expertise and human resources for practical consultation. In recent years, FRI researchers have cooperated in the aforementioned policy programmes and been engaged in building partnerships with other organisations. Furthermore, researchers have begun promoting the concept of Satoumi. The FRI's participation has thus bridged the gap in resource inventory and practical consultation in coastal communities.

In this study, we demonstrate how the FRI has introduced participatory methods for interacting with local communities. To determine a path that leads to ecosystem restoration and furthermore to achieve goals associated with Satoumi, we identified issues in local marine resources use and implemented improvement actions based on a bottom-up approach. Moreover, we established systematic observation of marine resources and their economic use. The promotion of the Satoumi concept relies not only on official funds and policy support but also entails the consideration of local characteristics, the demands of residents, and cooperation and partnerships among stakeholders in the area. Most importantly, enhancing local communities' capacity via environmental education and strengthening community resilience, as well as protecting local traditional cultures, are pivotal tasks. Thus, the objectives of the project were: (1) to integrate the different voices of multiple stakeholders to achieve harmony within specific social-ecological systems in coastal areas; (2) to empower local communities and other stakeholders to facilitate management in the SEPLS involving coastal villages; and (3) to engage local people in restoring ecosystems in the coastal SEPLS by developing a collaborative governance mechanism.

11.2   Methodology

The FRI adopted participatory action research (PAR) to establish a public–private collaboration framework for managing Mao'ao Bay's coastal SEPLS. To date, actions have been comprised of three main activities, which are discussed in the following sections.

This study used PAR to explore how Mao'ao Bay has practiced Satoumi and to establish a governance model for the SEPLS. PAR is a methodology for conducting actions to understand and improve an environment (Bennett 2004). This reflective process is directly linked to action, influenced by an understanding of history, culture, and local context, and is embedded in social relationships. The spirit of PAR allows the people taking part in reflective actions to have much more control over decisions that affect their own lives (Baum and Shipilov 2006).

We followed Bennett (2004) to conduct our PAR. First, the research team employed field surveys, secondary data analysis, questionnaire surveys, and in-depth interviews to collect basic information about Mao'ao Bay. We then focused on power relations and tried to blur their boundaries by holding stakeholder and

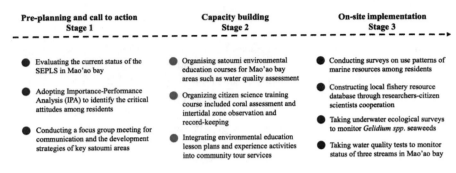

Fig. 11.2 Procedures of PAR (Source: Prepared by authors)

focus group meetings to allow participants to share views and co-create a Satoumi vision and strategy. The second stage focused on building partnerships with local people and empowering them by holding workshops, empowerment courses, and training sessions in environmental monitoring and underwater ecological surveys to facilitate management of the Mao'ao Bay SEPLS. In the final stage, the objective was to form a collaborative governance mechanism by engaging local people to play key roles in citizen science to restore the coastal ecosystem (Fig. 11.2).

11.2.1 Pre-Planning and Call to Action

We first implemented a questionnaire survey technique to evaluate the current status of the Mao'ao Bay SEPLS and assess nearby Mao'ao communities. We adopted importance-performance analysis (IPA) (Martilla and James 1977), which allowed us to identify the critical attitudes among residents to analyse the current state of Mao'ao Bay. This method is broadly adopted for the development of marketing strategies and combines respondents' perceived importance and performance of attributes into a two-dimensional plot to facilitate data interpretation. Attributes are classified into four quadrants to set priorities. Typically, the four quadrants are identified as "keep up the good work" (QI), "possible overkill" (QII), "low priority" (QIII), and "concentrate here" (QIV). The questionnaire design was based on the Japan Satoyama Satoumi Assessment (JSSA) indicators (JSSA 2010) and the Satoumi Manual of Japan (Ministry of the Environment of Japan 2011). From August to September 2020, IPA was conducted to evaluate residents' attitudes towards the promotion of Satoumi in Mao'ao Bay. Purposive and snowball sampling was employed to convene forty residents to participate in the questionnaire survey. They provided their perceptions on further strategies for the promotion of Satoumi.

We also conducted a focus group meeting to open multiple communication channels and discuss development strategies for Mao'ao Bay. In May 2020, one focus group meeting with 17 participants was held including two local residents, two officials from the fisheries sector in the New Taipei city government, six officials from the Gongliao District Office, one research fellow from the FRI, three researchers from National Taiwan Ocean University, and three representatives of a non-governmental organisation (NGO). The group was convened to explore the following overarching questions: "What are the strengths of and approaches to implementing Satoumi in Mao'ao Bay?" "How might these different approaches be applied to Mao'ao Bay?" and "What lessons can be learned from previous efforts to take action to implement the Satoumi approach?" The meeting provided an opportunity to guide the discussion to ascertain the needs and current situation of Mao'ao Bay and the community and to determine strategies for Satoumi development.

11.2.2 Capacity Building

At the capacity-building stage, we considered the empowerment of local communities and other stakeholders to facilitate management in Mao'ao Bay. Additionally, our purpose was to raise awareness among locals of marine environmental conditions, nurture relevant talent, recruit teams to promote Satoumi education, and build a human resources database.

In September 2020, after identifying and evaluating local social and ecological resources from field surveys, questionnaire surveys, and the focus group meeting, a critical issue was revealed—the community's concern for the Mao'ao Bay environment—especially for water quality and the resources on which residents relied. Thus, we first planned and organised Satoumi environmental education courses for the Mao'ao Bay area, such as on water quality assessment. This course educated local people on how to sample water from the three streams within Mao'ao Bay and how to test and record the data in cooperation with National Taiwan Ocean University. A total of 20 residents finished the course.

The other course was a citizen science training course designed to enable the locals to conduct intertidal/underwater ecological surveys and to encourage them to take action. The course included coral assessment and intertidal zone observation and record-keeping in cooperation with an NGO called the Taiwan Association for Marine Environment and Education and the local elementary school. The training course educated residents and local elementary students on monitoring the marine environment via adoption of scientific methods. A total of 29 participants finished the course.

11.2.3   On-Site Practical Implementation: Coastal Ecosystem Conservation

Citizen science has emerged as a way to generate large datasets and environmental awareness among target groups (Eitzel et al. 2017). Community-based monitoring (CBM) is a crucial element of many citizen science initiatives (Conrad and Hilchey 2011). CBM involves local stakeholders in monitoring local interests or concerns, ensuring citizens are involved as scientists, not just data collectors (Fulton et al. 2019). Thus, on-site practical implementation of the project included citizen science for the investigation of important native species such as *Gelidium* spp. seaweeds and other species. Actions taken at this stage were to engage local people in restoring coastal ecosystems in the SEPLS by developing a co-management mechanism.

In order to fully comprehend marine resource utiliation and its current status in Mao'ao Bay, FRI researchers cooperated with residents to collect logbook data on catch per unit of effort (CPUE), income per unit of effort (IPUE), harvested geographic location, and local perceptions of and attitudes towards marine resource use patterns assessed during the major coastal fishing season (April to October 2021). A total of 366 data points were collected from five residents with experience in harvesting ranging from 3 to 62 years. The collection of logbook data was performed to help us not only to observe how residents depend on the local fishery resources, but also to establish a record of harvested species and the socio-economic contribution of the resources. We also conducted workshops in the summer of 2021 with the Amas (people who harvest coastal marine resources by free diving) to identify relevant species from oral interviews and carry out underwater Satoumi actions (such as the citizen science capacity building for diving and water quality monitoring by residents). These actions enabled us to better grasp information on the daily lives of residents and the harvest to apply to the development path of the Satoumi.

Finally, to verify the interactive effects of seaweed production with different environmental factors or human disturbances, we cooperated with eco-scientist divers and residents to co-establish ecosystem data for Mao'ao Bay. For the early planning and implementation, this science-based action began with a focus on algal flora, which has both ecosystem and socio-economic value.

11.3   Results

11.3.1   Observations

11.3.1.1   Mapping Residents' Attitudes Through IPA

To establish the Satoumi strategies, we mainly focused on Quadrant IV, "Concentrate Here", as the points in this area have high priority for improvement and

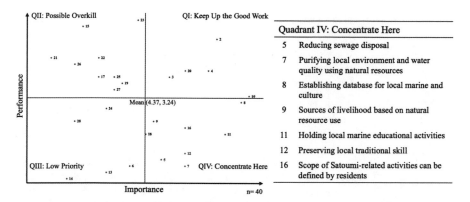

Fig. 11.3   The IPA results (Source: Prepared by authors)

opportunity, as perceived by residents (Fig. 11.3). There were seven items within Quadrant IV that required attention: residents assisting in reducing sewage disposal; use of natural resources to purify the local environment and water quality; establishment of database of local marine resources and culture; residents can earn living from natural resources; holding local marine educational activities; preservation of local traditional skill; and locals can define the scope of Satoumi-related activities (Fig. 11.3). Among these items, two main critical issues were identified. First, residents cared about sewage disposal but felt dissatisfied with the current situation. Second, residents thought highly about educational activities, the database for marine resources and culture, the benefits that locals could earn from nature, and the preservation of traditional skills. However, they were again found to be dissatisfied with the current situation of these issues. These attitudes and critical issues identified through IPA suggested a potential direction for Mao'ao Bay to implement Satoumi actions.

11.3.1.2   Developing Strategies Through a Focus Group Meeting

The focus group meeting in 2020 successfully created a space for stakeholders to discuss the future development of Mao'ao Bay. The four main issues identified were: "Collaborative Governance", "Resource Utilisation", "Cultural Inheritance", and "Citizen Science" (Table 11.2). These issues were then turned into actions to implement the Satoumi approach in Mao'ao. Furthermore, the meeting contributed to the establishment of the "Mao'ao Satoumi Platform". In 2021, the Platform played a critical role in allowing local people and other participants to brainstorm Satoumi actions for Mao'ao Bay. The platform held two meetings with a total of 64 participants in 2021. A consensus was reached among most participants on the importance of underwater ecological surveys and environmental monitoring.

Table 11.2 Summary of key messages collected at focus group meeting

Issues	Key messages
Collaborative governance	It is essential to obtain support from the local community, fishermen's association, central and local government, and academia/research to facilitate a governance network, which can encourage the participation of stakeholders.
Resource utilisation	Mao'ao endeavours to develop tourism. The participants are looking forward to the implementation of Satoumi, which can drive the transformation of traditional fisheries into recreational fisheries, the development of fishery experience tours, intertidal eco-tours, and promotion of fishery education.
Cultural inheritance	One of the critical tasks of Satoumi implementation is to pass down the local culture. The community started considering how to nurture talent, create innovative products, and establish promotional spaces. Thus, the stories of Mao'ao Bay and its fishing techniques could be inherited. The culture of the Mao'ao village and the community industry can be developed sustainably.
Citizen science	When facing the decreasing yields of important seaweeds, participants considered introducing citizen science to systematically investigate fishery resources and monitor other marine resources. Moreover, the collected data would be beneficial to the development of recreational fisheries and eco-tourism.

Source: Prepared by authors

11.3.2 Capacity Building

11.3.2.1 Enhancing Locals' Awareness of Resource Utilisation

During the water quality assessment courses, residents learned the technique to conduct water quality tests using API water test kits.[1] This helped them to understand anthropogenic impacts on their environment. For example, one of the streams in the Mao'ao Bay that was tested showed a higher value of NH_3-N that might be caused by residential sewage. Moreover, some of the participants mentioned that increasing sewage, discharged by restaurants, may also be affecting water quality in Mao'ao Bay (Fig. 11.4).

The citizen science training courses taught residents and local elementary students to monitor the marine environment by adopting scientific methods. The content of the course was designed according to the actual environment of Mao'ao Bay, for example, the characteristics of its marine ecosystem (e.g. coral reefs, sandy shores, intertidal zones, marine species and their characteristics, symbiotic relationships, food chain, and relevant research). Moreover, safety instructions and survey techniques on intertidal observation and underwater ecological monitoring were provided. After the training courses, two residents both with over 50 years of

---

[1] Including test kits for testing Ammonia (NH_3/NH_4), Nitrate (NO_3^-), Nitrite (NO_2^-), Phosphate (PO_4^-), and Carbonate Hardness (KH).

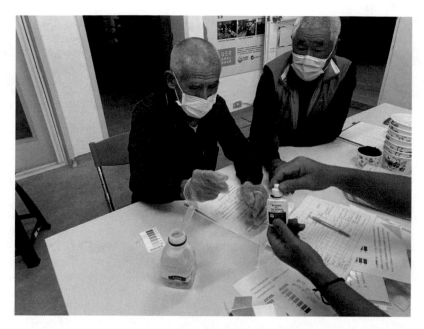

Fig. 11.4 Course on water quality assessment in the Mao'ao community (Photo taken by authors)

Fig. 11.5 Underwater ecological monitoring after citizen science training courses (Photos taken by authors)

experience in free diving in Mao'ao Bay engaged in co-establishing ecosystem data for Mao'ao Bay (Fig. 11.5).

| | | | |
| Anthocidaris crassispina | Gelidium elegans | Tripneustes gratilla | Eucheuma serra |

Fig. 11.6 Commonly harvested species in Mao'ao Bay (Photos taken by authors)

11.3.2.2 Transforming Experiences into Knowledge Facilitation

In order to fulfil the expectations of participants on developing tourism, the community integrated the Satoumi environmental education lesson plan[2] and experience activities (e.g. intertidal observation) into community tour services. For example, at present, the Mao'ao community offers environmental interpretation services and sampling of local traditional cuisine for visitors. Development of tourism activities offered a beneficial opportunity for the Mao'ao community to develop a model comprising both environmental education and industrial development. Furthermore, in our experiences in Mao'ao communities, we observed that local awareness and organisational capacity were increasing. The locals are now able to observe marine-related issues autonomously and have begun to interact with research institutes. The ongoing work in Mao'ao Bay and the experiences we have obtained constitute a practical methodology for the expansion of Satoumi in Chinese Taipei.

11.3.3 Coastal Ecosystem Conservation Actions

11.3.3.1 Local Marine Resource Utilisation Survey

Use patterns of marine resources in Mao'ao Bay obtained from logbook data showed a total of 2303.4 kg of catch (wet weight) in the coastal area of Mao'ao Bay by dive fishing or harvesting from April to October 2021. Of this, seaweed accounted for 70%, while the other 30% was high-value sea urchins and sea snails (Figs. 11.6 and 11.7). The harvest brought a total income of approximately 560,000 NTD[3] to five residents. The average monthly income per individual from April to September ranged from 5460 NTD to 28,760 NTD (approximately USD182–959) (Fig. 11.8). Regarding monthly average revenue, July was the highest (23% of total revenue), and October (4%) was the lowest, which indicates summer is the most important

---

[2]The Satoumi environmental education lesson plan was compiled after the capacity building stage and included in tour design.

[3]1 NTD (New Taiwan Dollar) equivalent to USD0.033.

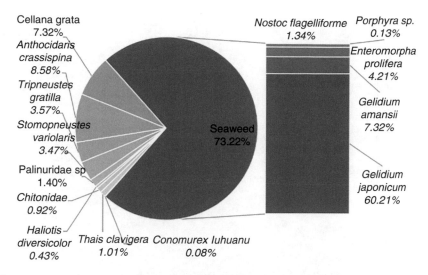

Fig. 11.7 2021 ratio of species harvested in Mao'ao Bay (Source: Prepared by authors)

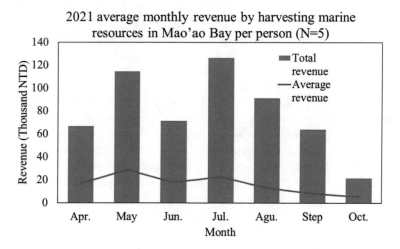

Fig. 11.8 2021 average monthly revenue per person by harvesting marine resources (Source: Prepared by authors)

harvest season. This highlights the importance of artisanal harvesting and fisheries in Mao'ao for livelihood support. Likewise, as residents also utilise the catch for their own consumption, it also is a critical source of nutrition for self-sufficiency.

The average monthly effort per resident ranged from 9.68 to 87.39 h, with the lowest monthly effort in April (15.8 h/month). In June, the average collection effort decreased to 17.67 h/month due to the COVID-19 pandemic. Amid the pandemic, residents greatly reduced the number of times they went out to harvest to maintain

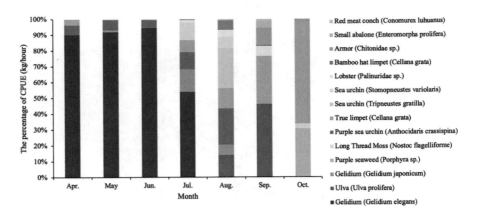

Fig. 11.9 Catch per unit of effort in the coastal areas of Mao'ao Bay by species per month (Source: Prepared by authors)

social distance from other people. Numbers rebounded to an average of 57.87 h per resident in August, 48.62 h in September, and 44.45 h in October.

The logbook data on harvests indicate that *Gelidium* is an economically important species in spring and summer. In the earlier harvest season, *Gelidium amansii* is the major harvested species due to its better gelatinisation, while *Gelidium japonicum*, commonly known as "dabinzai", is primarily harvested after August. This data shows the irreplaceability of *Gelidium* for supporting the livelihood and culture of the residents of Mao'ao Bay during all harvest seasons (Fig. 11.9).

The highest contributing species to the economy of the community by year were seaweeds, such as *Gelidium* and *Ulva*, followed by a sea urchin, mainly purple sea urchin (*Anthocidaris crassispina*) in the early season, which is replaced by *Tripneustes gratilla* and *Stomopneustes variolaris* after June. Lobsters have the highest catch value per unit effort, however, during a limited period (Fig. 11.10). A stable IPUE shown for *Cellana grata*, which is collected both for self-consumption and for giving to friends and relatives, demonstrates the cultural and social value of self-sufficiency, sharing, and giving in this fishing village culture.

In terms of the geographical distribution of harvesting (Fig. 11.11), Mao'ao Bay is the geographic harvest centre. Harvesting extends to the coasts on the east and west sides of the bay from April to October. The northeast side of Mao'ao Bay (Magang intertidal zone) exhibits a higher frequency of harvesting throughout the year due to its wider hinterland. Concerning seasonal changes, the frequency of fishing in April, July, and August is concentrated on the inner side of Mao'ao Bay. The area from Sandiao Cape to the Magang intertidal zone has a high frequency of fishing throughout the year.

Regarding sales channels, most residents in the sample sold products by themselves before the COVID-19 pandemic, while some were making stable sales to restaurants. Concerning attitudes towards resource utilisation, residents gave high scores on satisfaction with each harvest trip, and their harvesting experiences were not correlated with the IPUE.

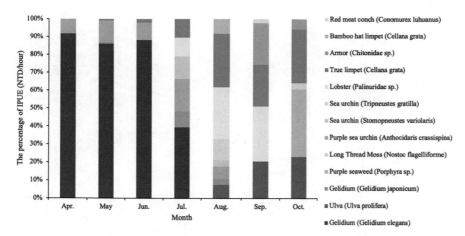

Fig. 11.10 Income per unit of effort in the coastal areas of Mao'ao Bay by species per month (Source: Prepared by authors)

The main motivation for the sampled residents to engage in harvesting, other than tidal factors, was based on the weather and sea conditions and related to their own physical condition. Specifically, collecting the coastal fishery resources not only provides economic support to residents, but is also part of their life. That is to say, the ecosystem services provided by the Mao'ao Bay are irreplaceable to the residents that harvest them.

Our study found that community households expressed an open attitude towards marine resource utilisation, believing that the ocean is public property and that everyone has the right to share its resources when the level of use is reasonable. However, they did express a desire to maintain the environment between the shore and the bay as tourist numbers increase. Local residents expressed their desire to demand respect for the local environment from visitors, such as by preventing them from leaving debris behind in the village and on the coast, and assuring tourists comply with regulations, such as parking their cars in the designated area.

11.3.3.2 New Actions: From Users to Citizen Scientists

From logbook data, we can identify information on important species for local residents. To construct a more systematic ecological survey, we expanded the findings through workshops held in 2021 for further discussion with residents. During the workshops, residents reported that *Eucheuma*, a type of seaweed that plays a very important economic and cultural roles in Mao'ao Bay, had become very scarce over the previous 2 years. Particularly in 2021, most residents reported that its production was very low, and only a few traces were found in the middle reef of Mao'ao Bay. Thus, due to insufficient production, residents were unable to sell

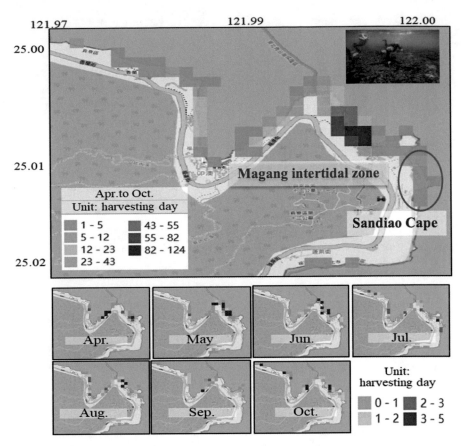

Fig. 11.11 Monthly geographical distribution of harvesting in Mao'ao Bay (Source: Prepared by authors)

Eucheuma during the spring and summer of 2021, with only a small amount collected by the Amas for their own consumption.

The residents believed that one of the reasons for the decreasing amount of *Eucheuma* was a change in the composition of nutrients and other substances in Mao'ao Bay caused by an increase in tourism and more sewage from restaurants. Accordingly, the residents applied what they learned in the environmental education courses to conduct water quality monitoring autonomously. Another factor noted by the residents that could explain the rapid decrease in *Eucheuma* production is the absence of typhoons (Hung et al. 2013), which could cause other algae to occupy the original habitat of *Eucheuma*. Through these activities, we observed that residents had not only gained the skills to monitor water quality and observe coral conditions, but also had a deeper understanding of the environmental impact of their own behaviour. This increased capacity and awareness of the residents shown in their activities illustrate how local people conducting scientific data collection can serve as a capacity-building exercise in future local promotion of citizen science (Freiwald et al. 2018).

11.3.3.3 Co-Creating Panel Data of Ecosystems

To figure out the disappearing *Eucheuma* phenomena, the last piece of the puzzle to be put in place was verification of the interactive effects of seaweed production with different environmental factors and/or human disturbances. A total of four scientific diving surveys were conducted in the spring (March to May) and summer (June to August) of 2021, at six sample sites (Fig. 11.12). The major purpose was to identify the dominant algae species and their seasonal changes in Mao'ao Bay.

A total of 63 species of algae were recorded in Mao'ao Bay during the year 2021 (Appendix, Table 11.3), including 9 species of *Chlorophyta*, 11 species of *Phaeophyta*, and 43 species of *Rhodophyta*. In May, 29 species were recorded at

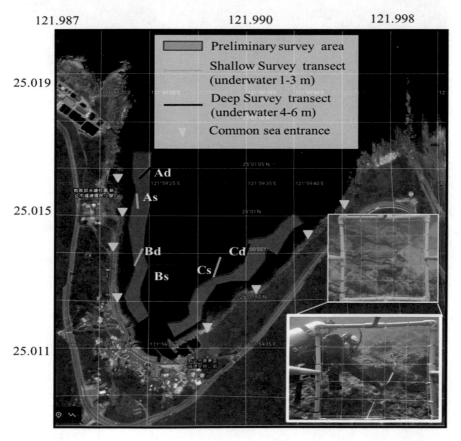

Fig. 11.12 Sampling sites of scientific diving research in Mao'ao Bay [Source: Reproduced from Google (2022); Embedded photos taken by authors]

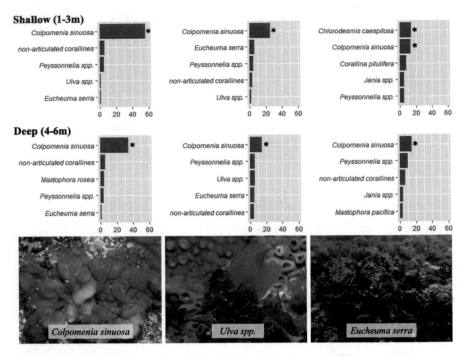

Fig. 11.13 Dominant algae species in May (Photos taken by authors)

As and 24 species at Ad, 36 species at Bs and 35 species at Bd, and 33 species at Cs, and 32 species at Cd. In August, the numbers were 23 species at As and 17 species at As, 22 species at Bs and 20 species at Bd, and 32 species at Cs and 21 species at Cd.

The number of species was higher in spring (May) than in summer (August). The algae coverage of Mao'ao Bay averaged 57.5% in May and 28.0% in August. The results indicate that location was the most important factor affecting the distribution of *Eucheuma*, followed by the depth and season (Fig. 11.13).

Areas with a high coverage rate of *Eucheuma* may be related to microhabitats. The areas with strong wave energy and more aerated water may meet the microhabitat conditions for the growth of *Eucheuma*. There could also be other factors, such as fewer predators (e.g. sea urchins or algal-feeding fishes), less fishing, or simply being located in the tail of a stream, where fragmentation of *Eucheuma* is suitable for reproduction.

We found no significant difference in depth among all sample stations regarding the natural distribution of *Eucheuma*, implying that the effects of fishing pressure, water temperature, and wave conditions at both shallow and deep sample site gradients were not significant this year. However, the two depths of 1–3 and 4–6 m may not significantly affect the natural distribution of *Eucheuma*. This means that if there are drastic changes in harvesting activities, water temperatures, and wave conditions, there is still an opportunity to compare these two depth differences at the same location. Therefore, future underwater surveys of *Eucheuma* should incorporate factors such as harvesting activity, water temperatures, and wave conditions. These results were to contribute to the experimental design of scientific diving surveys the following year in plans to centralise survey resources and make the survey more efficient. In addition, although there was no significant difference between the two depths for *Eucheuma*, other species such as the green algae *Chlorella Taiwan*, showed different distributions affected by microhabitat conditions. This suggests further research on environmental elements of the important microhabitat is required.

A seasonal effect on the natural distribution of *Eucheuma* was not significantly different among the sample stations, indicating that the coverage rate of *Eucheuma* did not decrease after the collection season (mid-March to the end of August) for the survey year. *Eucheuma* is perennial, so this year's coverage may affect the growth of the following years. However, seasons significantly affect the coverage of algae species and cause dominant species to change. Whether there is competition or growth of other algae must be further observed. Accordingly, further resource monitoring of *Eucheuma* needs to examine a single species while considering the status of other algae. In the future, we suggest that water temperature, water quality, harvesting activity, and competition should be included as factors in such monitoring.

11.4 Lessons Learned

To promote Satoumi in Chinese Taipei, the FRI aimed to identify a mutually beneficial relationship between humans and nature that not only focuses on environmental resources, but also encourages sustainable economic development and social networks in coastal areas. Thus far, the partnership has involved the public and private sectors and NGOs, and over 200 individuals have participated directly or indirectly in the process. Our study, therefore, yielded the following outcomes:

1. Having ascertained diverse attitudes and opinions among local stakeholders through the questionnaire survey and the focus group meeting, we suggested techniques to encourage interactions among stakeholders, facilitate consensus, and intensify partnerships between local and relevant organisations. In summary, the applied techniques proved useful for developing strategies for the advancement of Satoumi.
2. The introduction of Satoumi environmental education and citizen science training courses to residents in the communities increased their awareness of the condition of their marine environment. Furthermore, the content of these community empowerment courses was based on local awareness and the results of local resource inventories. This increased the willingness of community members to participate. The Satoumi concept and actions introduced in Mao'ao Bay were found to be effective for SEPLS management. Although the water quality and underwater monitoring are still a work in progress, these actions encouraged the residents to take autonomous precautionary measures. This may be helpful for

further preventing and halting destruction of coastal ecosystems and achieving restoration objectives in the future.

3. Providing practical methods for acquiring knowledge on marine resources and fishery management encouraged community members to continuously promote Satoumi actions. Additionally, the Satoumi education lesson plans offered an opportunity for the Mao'ao community to promote its fishing culture, local ecological knowledge, and related ecosystem services.

4. The involvement of multiple stakeholders in community-based resource restoration in Mao'ao Bay formed the actor network for collaborative governance. Such multi-stakeholder collaboration establishes multiple communication channels that facilitate the exchange of techniques and information for the purpose of resource management and economic development. Moreover, we found that this governance network for the SEPLS in Mao'ao Bay enhanced stakeholders' sense of identity and their willingness to engage in restoration actions.

5. We have established a "Satoumi Information Platform" for broader dissemination of the Satoumi concept and case studies (see https://satoumi.tw) to draw public attention to coastal areas. The platform also serves as a forum to facilitate the cooperation of cross-governmental bodies to promote Satoumi activities in more comprehensive and diverse ways. The ongoing task is to recruit local people with multi-disciplinary talent into our "Satoumi Extension Task Force" to enrich human resources and to seek more potential and creative ways to engage in transdisciplinary cooperation for restoration actions.

11.5   Conclusions

This study demonstrated the practice of the concept of Satoumi in Chinese Taipei's coastal fishing community, aiming to improve the well-being of coastal people and the resilience of a coastal SEPLS. Through surveys on the utilisation of local fishery resources and species, as well as consultations with residents on development, we sought strategies to improve the well-being of Mao'ao Bay. We attempted to enhance productivity and biodiversity through human manipulation and interaction between humans and nature, specifically by enrolling the residents in training courses. Therefore, coastal people were found to play a key role in restoring coastal ecosystems, not only as users and observers, but also as recorders and conservation practitioners. The relationships between the various fishery resources used by coastal people and their marine habitats can be further enhanced by empowering local people. This could help local people to understand how dynamic systems for implementing restoration measures can be developed, and to consider restoration methods and effectiveness in a holistic manner in response to natural conditions and human activities. Therefore, we realised the importance of improving our understanding on the connection between important species and their habitats through scientific surveys with the participation of the residents themselves.

Local people are now partly involved in Satoumi actions. This local involvement has played a vital part in the long-term coastal ecosystem monitoring and restoration in Mao'ao Bay. We believe that local people, scientists, and other stakeholders could cooperate to govern their coastal ecosystems and their communities in the short term. Finally, based on our practical experience, we suggest that the participatory methods we employed could be inspiring to other cases within a specific context, especially in coastal areas.

Acknowledgements Sincere gratitude and thanks are expressed to residents, participants, researchers, and other stakeholder groups of the Mao'ao community for providing valuable assistance for the study. Certain contents in this study have been submitted to the IPSI Secretariat as an IPSI case study. The work was funded by the "Establishment of a National Ecological Conservation Green Network" policy programme of the Council of Agriculture, Executive Yuan, R.O.C. (Chinese Taipei).

Appendix

Table 11.3 The composition and coverage of algal species in Mao'ao according to the 2021 scientific diving survey

Chlorophyta	Phaeophyta	Rhodophyta
Ulva fasciata	Colpomenia sinuosa	Tricleocarpa fragilis
Ulva pertusa	Dictyopteris repens	Liagora ceranoides
Ulva spp.	Dictyota friabilis	Dichotomaria marginata
Ventricaria ventricosa	Dictyota bartayresinana	Galaxaura spp.
Bryopsis plumosa	Dictyota patens	Yonagunia taiwani-borealis
Caulerpa chemnitzia	Dictyota dichotoma	Yonagunia palmata
Chlorodesmis caespitosa	Dictyota sp1	Polyopes polyideoides
Codium mamillosum	Padina spp.	**Gelidium elegans**
Chaetomorpha spiralis	Lobophora variegata	Pterocladiella capillacea
	Zonaria diesingiana	Dudresnaya japonica
	Spatoglossum stipitatum	Rhodopeltis borealis
		Meristotheca papulosa
		Chondrus ocellatus
Photos of the algal species		Chondrus verrucosa
		non-articulated corallines
		Mastophora rosea
		Mastophora pacifica
		Corallina pilulifera
		Corallina aberrans
		Corallina crassissmum
		Corallina maxima
Ulva fasciata	Gelidium elegans	Jania spp.
		Amphiroa spp.

(continued)

Table 11.3 (continued)

Chlorophyta	Phaeophyta	Rhodophyta
		Eucheuma serra
		Sarcodia suiae
		Peyssonnelia distenta
		Sonderophycus caulifera
		Peyssonnelia conchicola
	Yonagunia taiwani-borealis	*Peyssonnelia boergesenii*
Eucheuma serra		*Peyssonnelia* spp.
		***Gracilaria* spp.**
		Champia bifida
		Ceratodictyon repens
		Grateloupia spp.
		Grateloupia ramosissima
		Hypnea charoides
		Hypnea pannosa
		Chondria armata
		Chondria ryukyuensis
Colpomenia sinuosa	*Chlorodesmis caespitosa*	*Portieria hornemannii*
		Laurencia dendroidea
		Laurencia brongniartii
		Laurencia spp.

Photos taken by authors
Note: Algae species with bold represent those with common economic uses

References

Baum JA, Shipilov AV (2006) Ecological approaches to organizations. In: Clegg SR, Hardy C, Lawrence T, Nord WR (eds) The SAGE handbook of organization studies. Sage, London

Bennett M (2004) A review of the literature on the benefits and drawbacks of participatory action research. First Peoples Child Fam Rev 1(1):19–32

Chen JL, Hsu K, Chuang CT (2020) How do fishery resources enhance the development of coastal fishing communities: lessons learned from a community-based sea farming project in Taiwan. Ocean Coast Manag 184:105015

Conrad CC, Hilchey KG (2011) A review of citizen science and community-based environmental monitoring: issues and opportunities. Environ Monit Assess 176(1):273–291

Eitzel MV, Cappadonna JL, Santos-Lang C, Duerr RE, Virapongse A, West SE et al (2017) Citizen science terminology matters: exploring key terms. Citiz Sci Theory Pract 2(1):1. https://doi.org/10.5334/cstp.96

Freiwald J, Meyer R, Caselle JE, Blanchette CA, Hovel K, Neilson D, Dugan J, Altstatt J, Nielsen K, Bursek J (2018) Citizen science monitoring of marine protected areas: case studies and recommendations for integration into monitoring programs. Mar Ecol 39:e12470. https://doi.org/10.1111/maec.12470

Fulton S, López-Sagástegui C, Weaver AH, Fitzmaurice-Cahluni F, Galindo C, Fernández-Rivera Melo F et al (2019) Untapped potential of citizen science in Mexican small-scale fisheries. Front Mar Sci 6:517

Gain AK, Ashik-Ur-Rahman M, Vafeidis A (2019) Exploring human-nature interaction on the coastal floodplain in the Ganges-Brahmaputra delta through the lens of Ostrom's social-ecological systems framework. Environ Res Commun 1(5):051003

Google (2022) Google earth [Mao'ao Bay, New Taipei City, Taiwan]. https://www.google.com/intl/zh-TW/earth/. Accessed 3 Jan 2022

Huang H-W, Chuang C-T (2010) Fishing capacity management in Taiwan: experiences and prospects. Mar Policy 34(1):70–76

Hung CC, Chung CC, Gong GC, Jan S, Tsai Y, Chen KS et al (2013) Nutrient supply in the southern East China Sea after typhoon Morakot. J Mar Res 71(1–2):133–149

Japan Satoyama Satoumi Assessment (JSSA) (2010) Satoyama-Satoumi ecosystems and human well-being: socio-ecological production landscapes of Japan—summary for decision makers. United Nations University, Tokyo

Lee KT (2015) On the promotion of demonstration area of fish farming in the coastal waters of Taiwan. Research Report of Fishery Agency, Council of Agriculture. (in Chinese)

Martilla JA, James JC (1977) Importance-performance analysis. J Mark 41(1):77–79

Ministry of the Environment of Japan (2011) Satoumi manual. https://www.env.go.jp/water/heisa/satoumi/common/satoumi_manual_all.pdf. Accessed 29 June 2022. (in Japanese)

Worm B, Hilborn R, Baum JK, Branch TA, Collie JS, Costello C, Fogarty MJ, Fulton EA, Hutchings JA, Jennings S (2009) Rebuilding global fisheries. Science 325(5940):578–585

Yanagi T (2013) Japanese commons in the coastal seas: how the Satoumi concept harmonizes human activity in coastal seas with high productivity and diversity. Springer, Netherlands

Zhou S, Smith AD, Punt AE, Richardson AJ, Gibbs M, Fulton EA, Pascoe S, Bulman C, Bayliss P, Sainsbury K (2010) Ecosystem-based fisheries management requires a change to the selective fishing philosophy. Proc Natl Acad Sci 107(21):9485–9489

The opinions expressed in this chapter are those of the author(s) and do not necessarily reflect the views of UNU-IAS, its Board of Directors, or the countries they represent.

Open Access This chapter is licenced under the terms of the Creative Commons Attribution-NonCommercial-ShareAlike 3.0 IGO licence (http://creativecommons.org/licenses/by-nc-sa/3.0/igo/), which permits any noncommercial use, sharing, adaptation, distribution and reproduction in any medium or format, as long as you give appropriate credit to UNU-IAS, provide a link to the Creative Commons licence and indicate if changes were made. If you remix, transform, or build upon this book or a part thereof, you must distribute your contributions under the same licence as the original. The use of the UNU-IAS name and logo, shall be subject to a separate written licence agreement between UNU-IAS and the user and is not authorised as part of this CC BY-NC-SA 3.0 IGO licence. Note that the link provided above includes additional terms and conditions of the licence.

The images or other third party material in this chapter are included in the chapter's Creative Commons licence, unless indicated otherwise in a credit line to the material. If material is not included in the chapter's Creative Commons licence and your intended use is not permitted by statutory regulation or exceeds the permitted use, you will need to obtain permission directly from the copyright holder.

Chapter 12
Multi-Stakeholder Platform for Coastal Ecosystem Restoration and Sustainable Livelihood in Sanniang Bay in Guangxi, South China

Yufen Chuang, Xin Song, and Guanqi Li

12.1 Introduction

12.1.1 Environmental and Socio-Economic Characteristics of the Area

Beibu Gulf is located in the northwestern part of the South China Sea and comprises the shallow water bay between China's Guangxi Zhuang Autonomous Region, Hainan Island, Leizhou Peninsula, and Vietnam. It lies in a tropical coastal landscape where the climate is humid and warm, and is classified as a northern tropical monsoon climate. The gulf is rich in ecological types and diverse in species. It is an important land and sea ecological barrier at the southern tip of China, and also an important global biodiversity hot spot. It is one of the key landscapes in the Country Programme Strategy of the GEF Small Grants Programme (SGP). Mangroves in Beibu Gulf are widely distributed, making it one of three major distribution areas for mangroves in China. It is also an important ecological zone for seagrass beds, coral reefs, and coastal wetlands.

12.1.2 Problem Analysis

Despite its ecological richness and diversity, Beibu Gulf has been faced with exploitation of its ecological resources and culture over the past 20 years alongside

Y. Chuang (✉) · X. Song · G. Li
Farmers' Seed Network China, Nanning City, Guangxi, China
e-mail: zhuangyufen@fsnchina.net

© The Author(s) 2023
M. Nishi, S. M. Subramanian (eds.), *Ecosystem Restoration through Managing Socio-Ecological Production Landscapes and Seascapes (SEPLS)*, Satoyama Initiative Thematic Review, https://doi.org/10.1007/978-981-99-1292-6_12

rapid development. Coastal erosion, sedimentation, pollution, degradation and reduction of wetlands in the intertidal zone, and decline in biodiversity are the main challenges faced by the region.

The cultivated land along the Beibu Gulf coast is fragmented. The proportion of agricultural income of villagers is low, while the proportion of fishing income is relatively high. Before the development of tourism, the main source of income for the locals was fishing, which was supplemented by agriculture. However, due to the deterioration of the marine ecological environment, fishery resources have been severely depleted. Income from fishing is at present much lower than it was previously. Some fishermen use electric trawls and other methods that destroy microorganisms in the ocean and seriously harm the ecological balance, worsening the situation.

Overfishing not only threatens the survival of Chinese white dolphins by affecting the food chain, but also seriously affects the livelihoods of local people who rely on fishing and tourism. Enclosed tideland cultivation has encroached on mangroves along the coast, affecting species in the intertidal zone and shallow sea fish, which has directly reduced the types and quantities of species that Chinese white dolphins prey on. As a result, the dolphins' range and movement patterns have gradually narrowed, and biodiversity and ecological functions have been lost. This has directly affected the marine ecology in which this wild and endangered species lives, as well as the diversity of income sources of the fishermen who have lived here for generations. According to local villagers, the population of white dolphins was relatively large in the past, and there has been a recent trend of decline. Most of the residents support the protection of white dolphins and believe that the presence of white dolphins indicates a healthy natural environment.

The recent urbanisation of coastal areas and the industrialisation of fisheries has brought challenges to traditional fishing communities. As offshore fishery resources continued to be depleted, traditional fishing communities face a decline. These communities also face various other challenges, including population ageing, few employment opportunities, serious hollowing of rural areas, and low awareness of environmental protection and planning among fishermen. Thus, there is an urgent need to improve their village landscapes (Wang et al. 2018).

Because of its rich fishery resources, Sanniang Bay, which lies in Beibu Gulf, is home to Endangered, Threatened, and Protected (ETP) species, including the *Chinese white dolphin* (*Sousa chinensis*), horseshoe crabs (*Tachypleus tridentatus* and *Carcinoscorpius rotundicauda*), and mangroves. The Chinese white dolphin is one of the species of wildlife protected at the national level in China, together with the dugong. Chinese white dolphins are listed as a species threatened with extinction in Appendix I of the Convention on International Trade in Endangered Species of Wild Fauna and Flora (CITES 2022), and were also listed as a vulnerable species in the IUCN Red List of threatened species in 2015 (Jefferson et al. 2017; Laurie et al. 2019).

In 2003, Sanniang Bay Village was designated as the Sanniang Bay Tourism Scenic Area by the Qinzhou municipal government. While some villagers have gained profit from tourism development, fishermen have faced a dilemma over livelihood transformation. Overfishing not only threatens the survival of dolphins, but also seriously affects the subsistence livelihood of locals.

The zoning of the proposed Sanniang Bay Chinese White Dolphin Nature Reserve and tourism development have created a conflict between conservation and development, making it a great challenge to balance the two. Many scientific research institutions in the area have already conducted a large number of scientific surveys and developed systematic conservation measures, accumulating a wealth of scientific data. But there has been insufficient communication with the villagers, who have little understanding of scientific conservation. Moreover, these assessments involved very limited participation from villagers. Likewise, government agencies are very familiar with the concept of "balancing environmental protection and sustainable utilisation of resources", but have limited knowledge of community-based conservation of natural resources in Indigenous Community Conserved Areas (ICCAs). In November 2012, the 18th Communist Party of China National Congress for the first time stressed the "high priority to making ecological progress and incorporate it into all aspects". Ever since, the concept of "ecological civilisation" and environmental protection have been promoted in an accelerated manner. China's policy shift to preserve ecology and the environment calls for conceptual and practical change to value a biodiversity-friendly fishery system. Hence, recognition of the importance of community participation in protecting marine resources is definitely needed by both scientists and government agencies.

12.1.3 Objectives and Rationale

According to the IUCN's definition, Indigenous and Community Conserved Areas (ICCAs) are "natural and/or modified ecosystems, containing significant biodiversity values, ecological benefits and cultural values, voluntarily conserved by indigenous peoples and local communities, through customary laws or other effective means" (Kothari et al. 2012, p. 155). Community participation has always been the cornerstone of community development, environmental protection, and sustainable ecotourism development.

This case study highlights an interdisciplinary and multi-stakeholder participatory approach to exploring a community-based conservation mechanism. The study aims to analyse the processes and outcomes of a pilot initiative aimed at the protection and sustainable development of fishing communities in the Beibu Gulf that took place for the period from January 2020 to June 2022. Demonstrating how the community's actions with multi-stakeholder involvement can be collectively developed to build a collaborative network for capacity building, the study shows how the initiative engaged in awareness raising on marine resource conservation by introducing a community-based landscape approach.

12.2   Methodology

12.2.1   Study Site

The pilot site is located in Qinzhou City on Sanniang Bay, which is defined as a restricted redline area of concentrated distribution of Chinese white dolphins in the "Guangxi Marine Ecological Red Line Demarcation Plan" (Guangxi Zhuang Autonomous Region of Oceanic Administration 2018). The target sea area is composed of the Chinese White Dolphin Nature Reserve in the Qinzhou City of Guangxi Province, namely the sea area of Sanniang Bay and the estuary of Dafeng and Nanliu rivers in the Beibu Gulf, east to 108°47′49.20″ and west to 108°52′18.37″, north to 21°36′16.56″ and south to 21°31′11.72″ (Fig. 12.1). Sanniang Bay Village is under the jurisdiction of Rhino Foot Town, and consists of five villager groups with a total

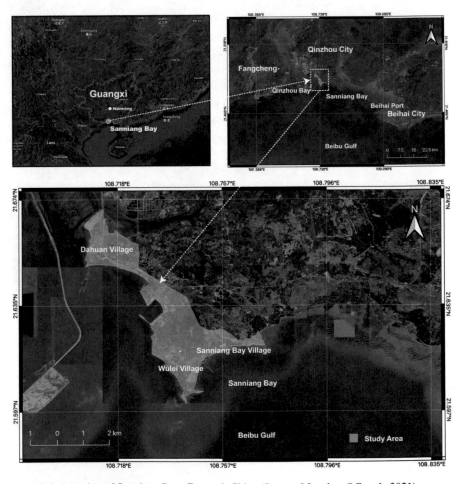

Fig. 12.1 Location of Sanniang Bay, Guangxi, China (Source: Map data ©Google 2021)

Table 12.1 Basic information of the study area

Country	China
Province	Guangxi
District	Qinnan
Municipality	Qinzhou
Size of geographical area (hectare)	418.85
Dominant ethnicity(ies), if appropriate	Han ethnic group
Size of case study/project area (hectare)	150
Dominant ethnicity in the project area	Han ethnic group
Number of direct beneficiaries (people)	2217 beneficiaries/520 households (Sanniang Bay Village)
Number of indirect beneficiaries (people)	44,385 (Rhino Foot Town)
Geographic coordinates (latitude, longitude)	21°37′17.07″N, 108°45′19.29″E

Source: Authors, based on the study

population of 2217 (Table 12.1). The Sanniang Bay is home to ETP species, including the Chinese white dolphin, horseshoe crabs, and mangroves.

According to information from the Sanniang Bay Tourism Management Zone Committee in 2019, the Sanniang Bay Village has a total area of 150 ha, of which 121 ha is slope land, and 2.5 km of coastline. The whole village lies within the Sanniang Bay Tourism Scenic Area. The villagers' livelihoods are dominated by marine fishing and supplemented by tourism and agriculture. There are 185 large and small motorised fishing boats. The fishermen's income depends on fish resources. During the fishing season, the daily income is not less than 300 CNY (equivalent to approximately USD44) per fisherman per day. Those villagers who do not have fishing boats work as sailors for those who do.

In 2003, Qinzhou City developed the Sanniang Bay Tourism Scenic Area. Consequently, villagers invested in the construction of restaurants and petty stores. Later, inns, retail stores, and tour guides joined to gradually form a service industry. In the village as a whole, about 60–70 people are engaged in the service industry, including catering, accommodation, sale of souvenirs, and white dolphin sightseeing services. Sanniang Bay Village won the award as the "most beautiful fishing village" in February 2018 by the Ministry of Agriculture and Rural Affairs.

12.2.2   Participatory Research Approach

Many conservation agencies along the Beibu Gulf coast focus on ETP species protection. However, fishing activities affect these protected species. To seek a sustainable solution, researchers initiated a pilot project to explore a community-based mechanism for balancing the habitat protection and development of fishing communities, employing an interdisciplinary and multi-stakeholder participatory approach.

12.2.3 Stakeholder Roles

The Farmers' Seed Network (FSN) was founded in 2013 in Guangxi, established on the basis of participatory action research conducted by multiple institutes including the Center for Chinese Agricultural Policy of the Chinese Academy of Sciences, the Chinese Academy of Agricultural Sciences, and the Maize Research Institute of the Guangxi Academy of Agricultural Sciences in Southwest China. FSN is a pioneering organisation in applying participatory research methods to examine agrobiodiversity and natural resource management in China.

In October 2018, FSN was officially registered as a non-governmental organisation in Nanning, Guangxi. As a non-profit social organisation, FSN has worked in over 40 rural communities in 10 provinces across the country to preserve plant genetic resources, support community-based seed conservation and sustainable utilisation, and facilitate productive collaborations between farmers and plant scientists for strengthening farmers' seed systems, improving farmers' livelihoods, enhancing farmers' dignity, and promoting national seed security.

Based on participatory research experience accumulated in the field of natural resource management, FSN initiated its first coastal participatory research to promote ICCAs in Sanniang Bay Village in 2019. This research is jointly supported by the Global Environment Facility (GEF) Small Grants Programme of the United Nations Development Programme (UNDP) and the Sowing Diversity = Harvesting Security (SD = HS) project of Oxfam Novib. The SD = HS project aims to support indigenous peoples and smallholder farmers in gaining the capacity to access, develop, and use plant genetic resources to improve their food and nutrition security under the conditions of climate change.

Building on the different roles of local communities, scientists, and NGOs in biodiversity conservation and utilisation (Table 12.2), FSN facilitates collaboration and cooperation between local communities, scientific communities, social organisations, and local governments (Fig. 12.2).

Table 12.2 Stakeholder representatives

Stakeholder groups	Representative groups
Local governments	– Sanniang Bay Tourism Management Zone Committee – Sanniang Bay Village Council
Communities	– Sanniang Bay Village (mainly), Wulei Village, and Dahuan Village
Local NGOs	– Guangxi Biodiversity Research and Conservation Association (BRC) – China Blue Sustainability Institute
Research Institutes/ Universities	– Guangxi University for Nationalities – Ocean College of Beibu Gulf University, Guangxi Key Laboratory of Beibu Gulf Marine Biodiversity Conservation – Guangxi Academy of Agricultural Sciences

Source: Authors, based on the study

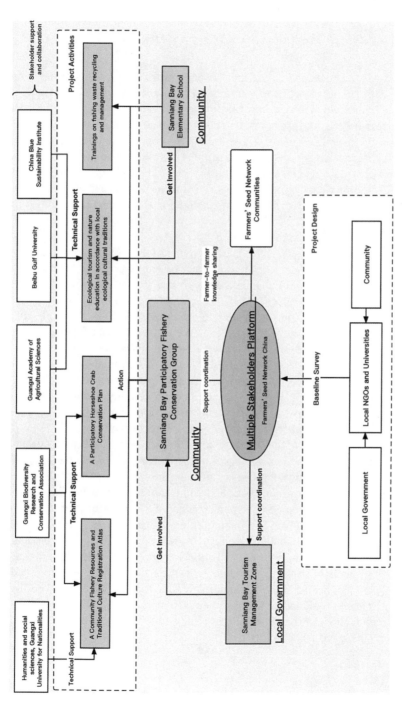

Fig. 12.2 Roles of stakeholders related to the project (Source: Authors, based on the study)

12.3   Description of Activities and Results

12.3.1   *Villager Participatory Fishery Conservation Group*

To gather baseline data and ensure that we address the actual needs of the community, we carried out a series of surveys and community engagement activities, including household surveys, key informant interviews, and group discussions. This process helped local participants and the project team to assess and reach a consensus on prioritised actions. The villagers proposed to use the community flexible fund provided by the project to support the establishment of a conservation group and provide a monthly labour subsidy of 400 CNY (equivalent to approximately USD58) per person for the villagers who patrol the beach every day.

The questionnaire surveys were conducted for the community from July to September 2020. A total of 33 locals participated. The collected data included socio-economic background information for persons from Sanniang Bay Village, Wulei Village, and Dahuan Village. The first round of the survey asked mainly about the status and changes in fisheries and the marine environment of the community. The second pre-survey was conducted through key informant interviews with the elders, focusing on the processes of the village development as well as the fishermen's transition from a collective economy (before the Reform and Opening-up in 1978) to the individual one. We also conducted focus-group interviews and community consultations between March and September 2021 to ascertain the perspectives of multiple stakeholders. Based on the above two rounds of baseline surveys and local research institutions' findings, the FSN project team realised that the Sanniang Bay Village lacks a community group. Organising a community-based participatory group would allow for the integration of community needs and conservation wants into planning and implementation.

After several rounds of interdisciplinary research team visits and continuous surveys, the project team identified several key persons willing to participate, and the Sanniang Bay Participatory Fishery Conservation Group was organised. The group consisted of ten core members, including the director of the Sanniang Bay Village Council, the principal and a young teacher of Sanniang Bay Primary School, fishermen, guesthouse owners, and local dolphin watch captains (Table 12.3).

Through monthly or bimonthly meetings, the project team invited researchers and social organisation staff to brief core members on ongoing research findings. We interpreted protection measures in the context of local practices and supported the villagers' action plans using the community flexible fund. Occasionally, we invited members to attend national workshops to engage in exchange with and learn about other successful local experiences in sustaining biodiversity and food systems in community-managed landscapes.

Table 12.3 Stages of the multi-stakeholder platform establishment and implementation

Stage	Approach	Activities and stakeholders
Identify stake-holders (July–September 2020)	*Two Pre-surveys*: Focus-group interviews, house-hold surveys, and key informant interviews	*Socio-economic survey and gauging level of interest in project participation* Stakeholders involved: – Local community – Local NGOs
Set up of multi-stakeholder plat-form (March–May 2021)	Community consultation and group meetings	*Organising a villager participatory fishery conservation group* (see Sect. 12.3.1) Stakeholders involved: – Local government – Tourism sector – Local community/Elders – Local school
Implementation (May–April 2022)	Daily catch monitoring and vil-lage transect walks	*Community Fishery Resources and Traditional Culture Registration Atlas* (see Sect. 12.3.2) Stakeholders involved: – Guangxi Biodiversity Research and Conservation Association (BRC) – Guangxi University for Nationalities
	Public participation and community-based monitoring	*Participatory Horseshoe Crab Conser-vation Plan* (see Sect. 12.3.3) Stakeholders involved: – Guangxi Key Laboratory of Beibu Gulf Marine Biodiversity Conservation, Beibu Gulf University
	Localised waste recycling and management training	*Training on recycling and management of beach waste and marine debris* (see Sect. 12.3.4) Stakeholders involved: – Local government – Tourism sector – Local community
	Establishing a pilot eco-guesthouse	*Exploring Alternative Livelihoods: Ecological Tourism* (see Sect. 12.3.5) Stakeholders involved: – Tourism sector – Guesthouse owners – Local dolphin watch captains
	Conservation education and public sensitisation	*Biocultural Education* (see Sect. 12.3.6) Stakeholders involved: – Local primary school – China Blue Sustainability Institute – Guangxi Academy of Agricultural Sciences

Source: Authors, based on the study

12.3.2 Community Fishery Resources and Traditional Culture Registration Atlas

In the past, the traditional fishing method used single-layer nets, which allowed smaller fish to escape. Recently, due to the development of pelagic fishery and fishing technology, single-layer nets have been transformed into three-layer nets to catch more fish. In response to increasing market demand, large fishing boats around Beibu Gulf are using high-density nets and other non-sustainable methods, gradually destroying the balance of marine ecological resources.

The local villagers of Sanniang Bay are still engaged in the traditional ecological fishing method. The majority of the fishermen in Sanniang Bay own family-run "husband and wife boats". Typically, couples head out to sea for fishing on small wooden fishing boats, sailing out in the morning and returning in the late afternoon (Fig. 12.3). Based on the pre-survey mentioned above, local fishermen have all noticed a decline in the types and quantities of local offshore catches.

In collaboration with the Guangxi Biodiversity Research and Conservation Association (BRC), a local NGO in Nanning, Guangxi, the project team benchmarked the Beibu Gulf marine species to generate a Sanniang Bay community fishery resources and traditional culture registration atlas and created a localised Daily Catch Monitoring Form for the participatory fishery conservation group (Fig. 12.4). The form

Fig. 12.3 Husband and wife boat (Photo taken by author)

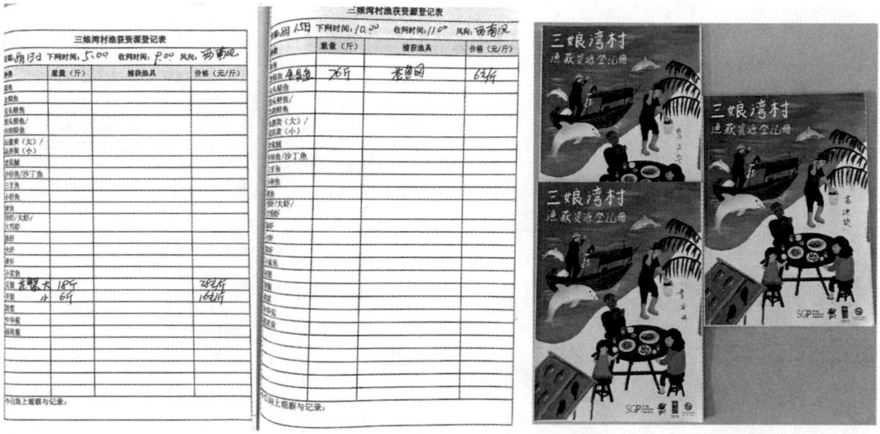

Fig. 12.4 Daily catch monitoring form (Photo taken by author)

was used by three group members to monitor their daily catches. We started with three fishing boats in an attempt to keep year-round records of fish species caught, weight, landed price obtained, and fishing methods used. Long-term monitoring data can show changes in common catch and market price fluctuations, which could help in further exploration of marine ecology and the fishermen's livelihoods.

To strengthen the connection between scientific knowledge and traditional practices, the project team carried out exchanges and cooperation with the Guangxi University for Nationalities and Beibu Gulf University to study and document local traditional knowledge in natural resources management.

12.3.2.1   Surveys on Fishing Culture and Documentation of Traditional Knowledge

In May 2021, participatory methods, such as group meetings and transect walks, were utilised to invite villagers, village cadres, and the Sanniang Bay Scenic Area Management Committee members to map out the community's natural and cultural resources (Fig. 12.5). These activities enabled the villagers to rethink and reconnect to the fishing village's value. To ascertain the perspectives of different stakeholders, the project team brought together the Sanniang Bay Scenic Area Management Committee, Sanniang Bay Village Council, fishermen, and guesthouse owners to illustrate the natural and social resources of Sanniang Bay. The process identified many local tangible and intangible cultural properties, ranging from historical buildings and dwellings to religious beliefs, family relations, handicrafts, folklore, and lifestyles. Results demonstrate how traditional ecological knowledge is embedded in the village.

Meanwhile, the project team also established cooperation with an anthropological research team at the Faculty of Humanities and Social Sciences of the Guangxi

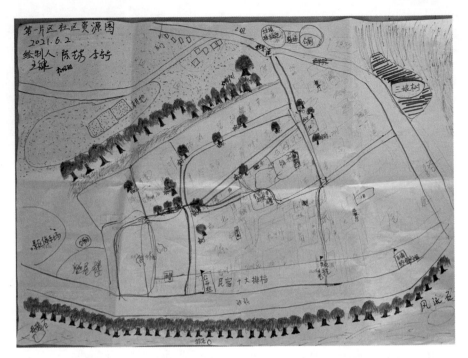

Fig. 12.5 Villagers map out the community's natural and cultural resources (Photo taken by author)

University for Nationalities to systematically study and analyse the local cultural context. This collaboration has supported the community's cultural and historical documentation and the formation of a solid foundation for the development of culturally appropriate ecotourism. The Sanniang Bay Scenic Area Management Committee also realised that aside from attracting tourists with the Chinese white dolphins, the development of the scenic area requires linkages with village culture. The fishermen and the sea are the heart and soul of the fishing culture. The sustainability of Sanniang Bay Village relies upon the inheritance of cultural heritage, honouring the harmony between humans and nature.

12.3.3   Participatory Horseshoe Crab Conservation Plan

Following the establishment of the Sanniang Bay Participatory Fishery Conservation Group, members raised several issues involving ETP species and local ecology protection. In addition to being the main habitat of the Chinese white dolphin, Sanniang Bay is also home to mangrove horseshoe crabs (*Carcinoscorpius rotundicauda*) and Chinese horseshoe crabs (*Tachypleus tridentatus*), both listed as Class II protected species (Laurie et al. 2019). On September 8, 2021, the

Fig. 12.6 Horseshoe crabs being released under the "Send Horseshoe Crabs Home" Initiative and the record-keeping template (Photo taken by author)

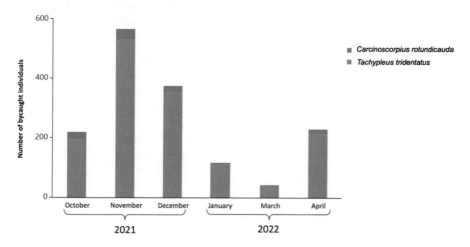

Fig. 12.7 Number of bycaught Chinese horseshoe crabs (*Tachypleus tridentatus*) and *Carcinoscorpius rotundicauda* in Sanniang Bay, Guangxi (Source: Guangxi Key Laboratory of Beibu Gulf Marine Biodiversity Conservation)

community group formulated an initiative called the *"Send Horseshoe Crabs Home"*. Five members rescue horseshoe crabs that have been accidentally caught in fishing nets and record their size and numbers daily (Fig. 12.6).

The record-keeping template was designed with help of a local research partner, Ocean College of Beibu Gulf University. Fishermen keep records of the width and number of horseshoe crabs during their daily work. Data obtained were provided to the Guangxi Key Laboratory of Beibu Gulf Marine Biodiversity Conservation to study the landing pathways of the horseshoe crabs (Fig. 12.7). This action is one in which local villagers can easily participate. The community group is planning to organise school children to join this initiative.

12.3.4 Training on Recycling and Management of Beach Waste and Marine Debris

According to the Sanniang Bay Tourism Management Zone Committee, the number of visitors to the bay has increased since entrance fees were waived in November 2020, and so has the amount of garbage. The management committee arranges garbage trucks and sanitation workers to clean the garbage drop-off sites twice a day during busy seasons.

Every year, from April to November, the monsoons affect the sea, and southwest winds blow garbage ashore. The beaches are often covered with household and marine garbage, which can be roughly separated into seashells, branches, leaves, plastic bottles, glass bottles, marine plastic, and nets. In the 1960s, local organic garbage such as seashells, with high calcium content, was crushed to feed chickens and ducks.

The management of marine debris must be performed as a continuous task to avoid garbage flowing back into the sea. Developing a circular economy by reusing sea waste is a possible solution. While it is important to deal with marine waste that has already been produced, it is even more critical to reduce marine pollution. Waste reduction and raising awareness on recycling are key elements of successful protection and management of marine ecosystems and their associated communities.

To bring localised waste recycling and management training to Sanniang Bay, the project team produced a short film called "Where the Trash Goes". This video documented the entire garbage collection, transfer, and disposal process in Sanniang Bay and the designated waste disposal facility. The film was shown to fifth graders at Sanniang Bay Primary School.

The project team also invited village council members to clean up the beach before sunrise. Through garbage classification and weighing of the debris, we saw how much hazardous and non-hazardous waste was on the beach and the surrounding village areas.

Working with the primary school, local artist Ba Nong and school kids turned the garbage debris collected from the mangrove forest into musical instruments and a giant "X" to serve as a stage background. An "X" plastic concert was held with music that incorporated traditional elements and brought the tradition to life, expressing children's belief in safeguarding the blue ocean. The children's performance at the beach attracted many tourists, who came to Sanniang Bay (Fig. 12.8).

Fig. 12.8 "X" plastic concert (Photo taken by author)

12.3.5 Exploring Alternative Livelihoods: Ecological Tourism

Since the Sanniang Bay Tourism Scenic Area was set up in 2003, it has benefited the community economically. Today, many tourists visit Sanniang Bay to see the Chinese white dolphins, without paying much attention to the fishers' lifestyle and village culture. Ecotourism has received much attention in recent years. We believe that the initial development of ecotourism must incorporate knowledge on local ecological resources and respect for nature's fragile equilibrium. Local villagers with knowledge of local flora and fauna, traditional culture, and values should be incorporated and consulted in the ecotourism planning and design stage. Doing so would not only reduce the impact of tourism on the natural ecology, but also increase a connection with local life and culture. Ecotourism can be seen as a tool to strengthen sustainable tourism because it emphasises sustainable tourism principles (Salman et al. 2020).

We promoted an environmentally friendly management model in local guesthouses. After participating in the community cultural documentation and species monitoring activities mentioned above, one of the group's guesthouse owners took the initiative to create an ecologically friendly service model, including: serving local ecological vegetables and fish caught by sustainable methods; reducing non-recyclable waste; and promoting friendly and non-disturbing dolphin-watching boats. In the near future, we plan to customise a local tour guide training programme. This will provide tourists coming to Sanniang Bay with a more in-depth and responsible travel experience. Concurrently, this programme aims to motivate villagers to learn more about the history and biocultural values of their village.

12.3.6   Biocultural Education

Sanniang Bay Tourist Scenic Area is targeted for the joint development of marine recreation and marine environmental protection. In addition to educating tourists on the concept of marine protection, residents in the community conservation areas also need courses on ecological conservation and management. According to the pre-survey and consultation with local schools, there is a lack of curriculum and educational resources based on local socio-ecological knowledge.

As needed, we invited marine biologists and nature art teachers to bring biocultural education and science popularisation classes into Sanniang Bay Primary School between March and September 2021. In the meantime, the China Blue Sustainability Institute launched a marine science education programme in the community, covering marine biodiversity, species conservation, and sustainable fisheries concepts.

Responding to the school's needs, we designed a series of activities for local children. In March 2021, we launched a reading room at the primary school. This reading room stocks more than 100 picture books and books on marine environmental protection. During the book shopping process, we realised there are very few marine conservation books targeting coastal village children in China. Therefore, we supported local teachers to develop a customised fishing culture curriculum, while inviting village elders to be lecturers. Meanwhile, we facilitated food education for children, setting up an eco-friendly vegetable garden and food education class at school. The school garden was popular among the students and teachers. In addition to local vegetable varieties, other community partners from the Farmers' Seed Network and the Academy of Agricultural Sciences provided indigenous seeds adapted to the Guangxi region (Fig. 12.9).

To make children cherish and care for the ocean around them, we also supported the formation of a "Sea Song Squad", with six children collecting local fishing songs, stories, and folklore.

12.4   Lessons Learned

After a series of explorations, a multi-stakeholder platform was established to improve coastal ecosystem restoration and sustainable livelihoods in Sanniang Bay. The initiative integrated scientific monitoring and baseline data collection into local fishermen's daily practices, as well as supported villagers to establish a participatory data collection mechanism and build consensus on conservation. Our findings are mainly reflected in the following areas:

(i) To raise adults' awareness of ecological conservation, an effective way is to work with primary school students, by setting up an eco-friendly vegetable garden, launching a marine-friendly reading room, and organising a beach clean-up concert.

Fig. 12.9 The eco-friendly vegetable garden at Sanniang Bay Primary School (Photo taken by author)

(ii) To explore alternative livelihoods for coastal communities, a feasible practice is to pilot community-led ecotourism combining community-guided tour services rooted in location-specific cultural elements, and a plastic-free, seed-to-table eco-guesthouse.

(iii) To mobilise individuals' and community participation, the optimal path is to facilitate exchange visits.

(iv) To establish an integrated beach waste management model in the Beibu Gulf, marine plastic debris should be monitored and sampled at a network of sites during the southwest monsoon season.

One of the highlights of this study's findings lies in the active participation of school children. The launch of the vegetable garden was a big hit among children, who were encouraged to grow, harvest, and cook their own food and share it with their families. This activity has enhanced the acceptance of the project team among the villagers. Another highlight is guaranteeing the community's participation with help of NGOs and scientists, since they are the ones who could provide accurate information on the social and environmental situation of the seascape. In particular, local NGOs played the intermediary role in capacity building, networking, and linkages for external support and exchange. Multiple-stakeholder platforms and networks with integrated approaches are crucial for supporting community-based conservation and development, as different stakeholders, including NGOs, scientists, civil society groups, enterprises, etc., can provide different information and support that communities need in the process at different times and in various situations. The experiences of Sanniang Bay Village are being scaled out to the two neighbouring communities of Wulei and Dahuan, as well as to Weizhou Island and Xia Village on the Beibu Gulf. The aim is to build a network to exchange lessons learned on biodiversity conservation and community development for further exploration of seascape-based ICCAs.

12.5 Conclusion

This case study especially highlights the significant role of community and smallholder fishermen in collaboration with multiple stakeholders on seascape restoration and community development. Despite facing challenges from biodiversity loss, industrial fisheries, and ocean degradation, the community strengthened their leadership and collaboration with social organisations, research institutions, and local governments via localised and culturally appropriate training, school education, and exchange visits. The Sanniang Bay fishing community's awareness and actions on conserving endangered species, including Chinese white dolphins, horseshoe crabs, and mangroves, and engaging in sustainable tourism and fisheries, were promoted via these initiatives.

We emphasise the importance of protecting the marine environment and coastal biodiversity, in addition to the protection of Chinese white dolphins and horseshoe crabs. We stress the key roles of communities and policy recognition for supporting community participation. We have illustrated community-based conservation as the main approach adopted by the case for involving multiple stakeholders in achieving sustainable development and financial security.

References

Convention on International Trade in Endangered Species of Wild Fauna and Flora, CITES (2022) Appendices I, II and III. https://cites.org/eng/app/appendices.php. Accessed 15 Aug 2022

Google (2021) Google maps of Qinzhou City, Guangxi, China. https://goo.gl/maps/asvqW8sr7Dk6NGAc6. Accessed 10 Dec 2021

Guangxi Zhuang Autonomous Region of Oceanic Administration (2018) The marine ecological red line was demarcated. http://hyj.gxzf.gov.cn/zwgk_66846/xxgk/fdzdgknr/bmwj/t3445428.shtml. Accessed 11 Nov 2021

Jefferson TA, Smith BD, Braulik GT, Perrin W (2017) Sousa chinensis (errata version published in 2018). The IUCN red list of threatened species 2017: e.T82031425A123794774. https://doi.org/10.2305/IUCN.UK.2017-3.RLTS.T82031425A50372332.en. Accessed 15 Aug 2022

Kothari A, Corrigan C, Jonas H, Neumann A, Shrumm H (eds) (2012) Recognising and supporting territories and areas conserved by Indigenous peoples and local communities: global overview and national case studies. Secretariat of the Convention on Biological Diversity, ICCA Consortium, Kalpavriksh, and Natural Justice, Montreal. Technical Series no 64, 160 pp. https://www.cbd.int/doc/publications/cbd-ts-64-en.pdf

Laurie K, Chen C-P, Cheung SG, Do V, Hsieh H, John A, Mohamad F, Seino S, Nishida S, Shin P, Yang M (2019) *Tachypleus tridentatus* (errata version published in 2019). The IUCN red list of threatened species 2019: e.T21309A149768986. https://doi.org/10.2305/IUCN.UK.2019-1.RLTS.T21309A149768986.en. Accessed 15 Aug 2022

Salman A, Jaafar M, Mohamad D (2020) A comprehensive review of the role of Ecotourism in sustainable tourism development. e-Rev Tour Res 18(2):215–233

Wang X, Wan R, Zhu Y (2018) Problems and countermeasures of China's fishery community development. Modern Agric Sci Technol 8:274–277

The opinions expressed in this chapter are those of the author(s) and do not necessarily reflect the views of UNU-IAS, its Board of Directors, or the countries they represent.

Open Access This chapter is licenced under the terms of the Creative Commons Attribution-NonCommercial-ShareAlike 3.0 IGO licence (http://creativecommons.org/licenses/by-nc-sa/3.0/igo/), which permits any noncommercial use, sharing, adaptation, distribution and reproduction in any medium or format, as long as you give appropriate credit to UNU-IAS, provide a link to the Creative Commons licence and indicate if changes were made. If you remix, transform, or build upon this book or a part thereof, you must distribute your contributions under the same licence as the original. The use of the UNU-IAS name and logo, shall be subject to a separate written licence agreement between UNU-IAS and the user and is not authorised as part of this CC BY-NC-SA 3.0 IGO licence. Note that the link provided above includes additional terms and conditions of the licence.

The images or other third party material in this chapter are included in the chapter's Creative Commons licence, unless indicated otherwise in a credit line to the material. If material is not included in the chapter's Creative Commons licence and your intended use is not permitted by statutory regulation or exceeds the permitted use, you will need to obtain permission directly from the copyright holder.

Chapter 13
Capacitating Philippine Indigenous and Local Institutions and Actualising Local Synergies on Restorative Ridge-to-Reef Biodiversity Conservation for Food Security and Livelihoods

Mark Edison R. Raquino, Marivic Pajaro, Jagger E. Enaje, Reymar B. Tercero, Teodoro G. Torio, and Paul Watts

13.1 Introduction

The institutionalisation strategy of the localised and programmatic restoration component is further validated through the use of resilience assessment. Restoration of socio-ecological production landscapes and seascapes (SEPLS) can be directly linked to the assessment of system resilience (UNU-IAS et al. 2014). In Aurora Province, Philippines (Fig. 13.1; Table 13.1), coastal food security and livelihoods are affected by localised actions on forest, agricultural, and marine ecosystems within the Sierra Madre Mountain Range (SMMR) and the North Philippine Sea (NPS) marine bioregion (Watts et al. 2021). The SMMR, a recognised biodiversity corridor, represents the country's longest mountain range that contains the largest remaining cover of old-growth tropical rainforest harbouring endemic and critically endangered species (Ong et al. 2002). The SMMR is home to some of the most ecocentric Indigenous Peoples and Local Communities (IPLCs) in the country. Aurora's globally significant rainforests at the centre of the SMMR are traditionally

M. E. R. Raquino · M. Pajaro · P. Watts (✉)
Daluhay Daloy ng Buhay Inc., Baler, Aurora, Philippines
e-mail: paulwatts@daluhay.org

J. E. Enaje
Bureau of Fisheries and Aquatic Resources Region 3, San Fernando, Pampanga, Philippines

R. B. Tercero
Agriculture Office, Provincial Government of Aurora, Baler, Aurora, Philippines

T. G. Torio
Environment and Natural Resources Office, Provincial Government of Aurora, Baler, Philippines

© The Author(s) 2023
M. Nishi, S. M. Subramanian (eds.), *Ecosystem Restoration through Managing Socio-Ecological Production Landscapes and Seascapes (SEPLS)*, Satoyama Initiative Thematic Review, https://doi.org/10.1007/978-981-99-1292-6_13

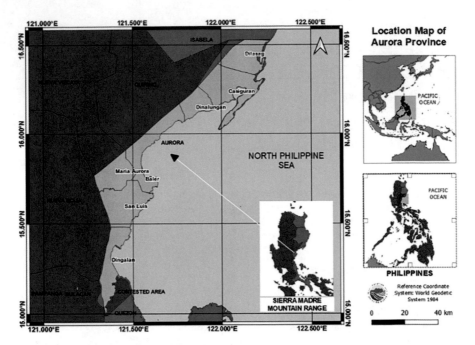

Fig. 13.1 Location map of Aurora Province within the Sierra Madre Biodiversity Corridor and the North Philippine Sea (Source: Authors, based on this study)

managed by Indigenous Dumagat, Alta, Agta, and Egongot communities that are determined to hold on to their culture, biodiversity heritage, and distinct ethnoecology. Marginalised fishing communities have settled along the coastline of Aurora; although primarily engaged in fishing, they are secondarily engaged in subsistence farming. These migrant settlers have now blended with the above-mentioned Indigenous communities, making up the IPLCs. It has been suggested that recognition of local institutions, knowledge, and tenurial rights produces effective conservation when it enables collective local environmental stewardship, which is dependent on social and political factors, including food security and support for livelihoods towards community resilience (Dawson et al. 2021).

This chapter considers reversing the trend of biodiversity decline through restoration of a globally significant area, as well as linked assessments and actions on both sociological and ecological factors. Critically, IPLC conservation efforts are harmonised with existing governance strategies by other local and national agencies. Several researchers have argued that IPLCs can be more than passive recipients of restoration activities as they have the historical reference and can therefore play more active roles in restoring ecosystems (Uprety et al. 2012). Direct involvement of IPLCs is perceived as important not only because it makes conservation more equitable, but also because it has the potential to produce more effective resource management and better biodiversity outcomes (Garnett et al. 2018). While this is the case for many IPLCs across the globe, legal and institutional empowerment and support for local institutions are still lacking. The pathways along which the role of

Table 13.1 Basic information of the study site

Features of the study area	Description
Country	Philippines
Province/State	Aurora
District	Region III
Municipality	Casiguran, Baler, San Luis, Maria Aurora and Dingalan
Size of the geographical area of analysis (hectare)	213,649 (Casiguran, Baler, San Luis, Maria Aurora and Dingalan)
Dominant ethnicity(ies), if appropriate	Dumagat, Agta, Alta and Egongot
Size of the case study/project area (hectare)	1. 140 ha of rainforestation of native trees in ancestral lands to support watersheds; 2. Establishment of 30-ha seagrass-mangrove protected area and 20-ha coastal greenbelts
Dominant ethnicity(ies) of the project area, if appropriate	Casiguran—Agta; San luis-Dumagat and Alta
Number of direct beneficiaries (people)	733
Number of indirect beneficiaries (people)	3000
Geographic coordinates (latitude, longitude)	Baler, capital of Aurora Province
Google map link for the study site	https://www.google.com/maps/d/u/0/edit?mid=1 ApAgCKr_PzG_PWTTiaHjQHuuD6SgOjyR&usp= sharing

Source: Authors, based on this study

IPLCs and characteristics of governance interact to produce different social and ecological outcomes have not been well explored, precluding a common understanding of these dynamics (Ferraro and Hanauer 2015). Harmonised governance strategies that prioritise IPLCs and empower them to contribute to long-term stewardship and effective conservation require approaches that enhance local capacity for management, planning, and governance as attributes of resilience.

The ridge-to-reef approach, including watersheds, agricultural lands, and marine ecosystems, encompasses interconnected social-ecological systems of natural and human resources taking into account the social, political, economic, and institutional factors affecting these systems (Katusiime and Schutt 2020). Watersheds descending from the SMMR irrigate thousands of hectares of life-giving farmlands, also providing both drinking water and hydroelectric power to the surrounding communities and provinces. These areas are a source of food, medicine, and livelihoods, as well as timber and non-timber products that help fill the demand for building materials, handicrafts, and furniture. The degradation of natural resources and biodiversity due to deforestation, conversion of the forest through unsustainable slash-and-burn farming techniques, wildlife hunting, charcoal making, unsustainable practices of harvesting timber and non-timber forest products, and poor governance can negatively affect dependent communities' food, health, and overall well-being (Golden et al. 2016; Pecl et al. 2017). Therefore, safeguarding and restoring ecosystems is often critical to support IPLCs' resilience and well-being (Sangha and Russell-Smith 2017).

IPLCs are considered as a focus of this case study, and where Indigenous groups are concerned, the study is based upon prioritising the goals of the United Nations Declaration on the Rights of Indigenous Peoples (UNDRIP 2013) that the Philippines has agreed to. Emphasis was also placed on activities that promote synergy and shared leadership between IPLCs as an approach to optimise community resilience. The pathways and processes presented in this case study further outline the ever-changing local SEPLS and how localised systems, including the IPLCs, deal with change and continue to develop. This chapter also recognises the significance of Indigenous and local knowledge (ILK) as critical information for restoring biodiversity, such as identifying restoration areas with the appropriate native species or cultural keystone places (Uprety et al. 2012; Cuerrier et al. 2015), both in terrestrial and in marine environments (Comberti et al. 2015). Further, ILK assists residents in withstanding shocks and disturbances and in using such events to catalyse renewal and innovation. However, enhanced community and ecosystem resilience are also important to meet increasing challenges facing ridge-to-reef ecosystem services and to ensure sustainable production systems. While these actions and goals are local, this case study is also intended to contribute to both global sustainable development objectives (UNU-IAS et al. 2014) and biodiversity targets (Tobón et al. 2017).

The objective of this chapter is to demonstrate how capacitating IPLC to implement a collaborative inter-community approach to institutionalised restoration activities can advance community-based resilience. The case study highlights the contribution of IPLC to biodiversity conservation and how inter-community synergies can optimise resilience and resources towards meeting larger spatial planning requirements (Strassburg et al. 2020) for ecological restoration. We also look into the complex pathways through which good governance qualities affect IPLC in the presence of enabling mechanisms and institutional factors that can empower stewardship actions with effective conservation outcomes.

13.2 Methods

The case study presented herein is based on a 2016 project implemented by the non-governmental organisation, Daluhay - Daloy ng Buhay, Inc. (Daluhay), on capacitation of Sierra Madre Indigenous and Artisanal Communities with support from the Global Environment Facility's Fifth Phase of the Small Grants Programme (SGP5). Seven IPLCs were selected as part of this project across five municipalities of Aurora Province. The organisations in the IPLCs were composed of two fisherfolk and five Indigenous communities from Egongot, Dumagat, Alta, and Agta communities. These communities expressed their interest in taking part in the initiative on community-based biodiversity conservation, having minimal experience in project implementation and financial management. For the Indigenous Peoples Organisations, a free, prior and informed consent was obtained from the Indigenous communities. The SGP5 initiative in Aurora Province concluded in 2018, but continuing initiatives after 2018 were considered in the results and discussion as some of the IPLCs scaled out their efforts.

13.2.1 Institutionalisation and Capacitating the IPLCs

The seven IPLCs (Fig. 13.2) were selected on the basis of having organised associations through other projects and being institutionalised by acquiring a legal personality through registration with a government entity such as the Department of

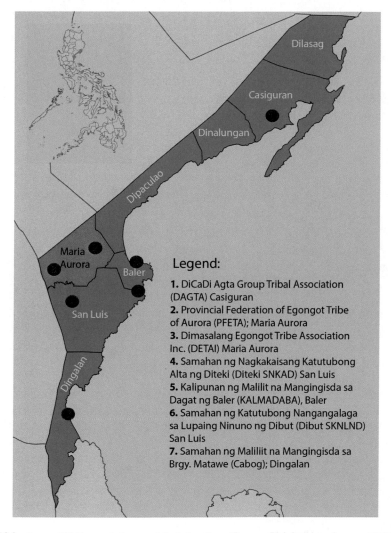

Fig. 13.2 Aurora IPLC associations and their locations (Source: Daluhay unpub. report 2018)

Table 13.2 Summary of components of institutional assessment

Component	Description
Governance	Organisation's accreditation/registration as a formal institution, presence of organisational structure and body
Administration	Presence and implementation of policies and systems in the organisation
Human Resources	Functionality of organisation based on membership roles
Financial Management	Presence of bank account and functionality of organisation's financial system, record keeping, monitoring, and audits
Operational Plan Development, Implementation and Networking	Functionality of organisation's operational plan related to biodiversity conservation and its capacity to establish linkages, support networks, and partnerships with other institutions

Source: Daluhay unpub. report (2018)

Labor and Employment or the Securities and Exchange Commission (SEC). Two of the seven IPLCs are fisherfolk organisations (i.e. KALMADABA and Cabog), while five are Indigenous Peoples Organisations, or IPOs (i.e. DAGTA, PFETA, DETAI, SNKAD, and SKNLND, see Fig. 13.2).

Capacity-building trainings and workshops were conducted for the selected community associations between 2016 and 2018 through SGP5. The trainings and workshops were focused on five components, which were later used in institutional and resource management effectiveness monitoring and evaluation (see Table 13.2).

Institutional Assessment and Analysis

An institutional assessment was conducted to determine the current capacity of the seven community associations. It was administered before and after the capacity-building interventions were conducted. The initial assessment conducted before the training served as the baseline from which the progress of the associations was tracked after subsequent interventions during SGP5 project implementation. The assessment was conducted through a focus group discussion with the officers and members of the associations, emphasising five key components: (a) governance, (b) administration, (c) human resources, (d) financial management, and (e) operational plan development, implementation and networking, as presented in Table 13.2.

The associations were assessed before and after the intervention using a Likert scale of 1–3, where 1 is lowest with little capacity, 2 middle, and 3 is the highest score. Depending on the components, several scenarios were stated in the assessment corresponding to the Likert scale. To assess any significant change brought about by capacity building, assessments were conducted for the organisations prior to trainings and again after the trainings. The Wilcoxon Signed Rank Test, which is a non-parametric statistical test that determines if two measurements of the same dependent variables have significantly changed given different conditions, was employed (Taheri and Hesamian 2013). To analyse the result of the assessment, the trial version of IBM SPSS software was used.

13.2.2 Actualising Local Synergies on Restorative Ridge-to-Reef Biodiversity Conservation for Food Security and Livelihoods

Ecosystem Restoration and Protection

Forest ecosystem restoration initiatives were carried out by all seven IPLCs through the community associations by planting native trees (i.e. rainforestation) within their degraded ancestral forests (for the Indigenous communities) or degraded mangrove and beach forest ecosystems (for the fishing communities). Utilising Indigenous or local knowledge, the IPLCs determined which native trees were to be outplanted in denuded forests, and wildlings were collected following trainings on nursery establishment and rain forestation techniques. The wildlings were potted and placed inside growth chambers for at least 3 months before outplanting in designated rain forestation areas. Monitors were assigned by each association on a rotation basis to check on the health of the wildlings and record mortalities in the growth chambers for replacement as necessary.

Linking different marine, agricultural, and forest ecosystems to help recover biodiversity was also undertaken via local government-declared and locally-managed protected areas as a form of other effective area-based conservation measures (OECMs). The OECMs in the IPLCs were initiated through the engagement of community associations, key local government officials, and the provincial Technical Working Group (TWG). Participatory resource assessments for coastal and agricultural ecosystems were conducted, after which community validation of the resource assessment results was performed. Later, a multi-stakeholder community working group was created to take the lead in the planning process for the formulation of a management plan for the locally managed network of protected areas.

Food Security and Sustainable Livelihood Development

The IPLC partners working on food security and sustainable livelihood development adhered to three elements in their initiatives: (1) ecological—with focus on conservation of biodiversity and sustainable use of biological resources; (2) economic—ensures that the enterprise is viable, sound, and sufficiently broad-based to create wealth, value, and positive returns benefiting the community, local government, and biodiversity; and (3) equitable—emphasising equal sharing and access to benefits by men, women, the Indigenous communities, and all other stakeholders in all economic activities. Whenever appropriate, the previously-mentioned participatory resource assessment results generated in relation to restoration efforts were utilised in the identification of sustainable livelihood interventions. As needed, relevant skills training for the identification of potential sustainable livelihoods and products (e.g. ecotourism, handicrafts, fish processing, and herbal medicine production) and formulation of project proposals and business plans were provided through the support of national government agencies and non-governmental organisations. Documentation of relevant local knowledge, skills, and practices was facilitated to support preservation of Indigenous culture and traditions through a skill transfer training held by elders and targeting the youth. This training was led by PFETA, whose aim was to provide guidance in establishing the Egongot Village as a potential eco-cultural tourism destination.

Networking and Building Linkages

To consolidate and scale up the SEPLS initiative in Aurora Province, Daluhay, with the community associations, shared best practice experiences and learnings during province-wide trainings and workshops. Daluhay also made efforts to build its directory of contacts, particularly with other civil society organisations, academe, local government units (both from within the province and neighbouring provinces and surrounding municipalities), potential funders, and national government agencies. These efforts are focused on scaling up the ridge-to-reef approach (Alejos et al. 2021) and professionalising Philippine Coastal Resource Management skills (Pajaro et al. 2022; 2013), within the biodiversity land corridor and marine bioregional considerations (Watts et al. 2021). These contacts were shared with partner IPLCs, providing them with opportunities to forge links with the network whenever relevant for them. These linkages are critical, particularly for policy advocacy and scaling up and scaling out of activities.

13.3   Results and Discussion

13.3.1   *Institutionalisation and Capacitating IPLCs*

The   Daluhay   facilitation   and   institutionalisation   paradigm   described in part herein, involves a reflexive focus on sustainability-gaps up through all levels of cultural mandate and stakeholders, project by project, often involving tens of "communities   of   interest"   within   specific   geographic   boundaries.   The institutionalisation phase of the current project included the process of establishing organised groups in IPLCs, was important as this is when the community associations were formed with the aim of being permanent and publicly recognised entities. The associations could then have a legal personality and begin to build their track record in project and financial management. Recognising the diversity of cultures with the selected IPLCs, Indigenous political structures needed to be considered in establishing organisational management systems in the case of the Indigenous communities. This initial step was greatly supported by the National Commission on Indigenous Peoples. In the case of the Egongot IPLC, clan representation needed to be ensured and considered in the selection of leaders.

While each community-based association in this study exhibited strong leadership and specific skills, emphasis was placed on building capacity through activities that not only strengthened the organisations themselves, but also promoted synergy and shared leadership between organisations. The trainings were held simultaneously for all IPLC partners, which optimised resources, but at the same time

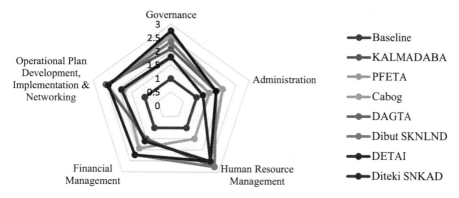

Fig. 13.3 Spider graph representation of mean values for the IPLC associations before and after capacity building (Source: Daluhay unpub. report 2018)

opened up opportunities for collaboration and synergy. For example, the trainings on environmental law enforcement led to a consolidated plan based on shared resources and coordinated actions among the IPLCs.

Institutional Assessment and Analysis
Involving multiple communities in training and activities led to associations and individuals taking leadership roles on specific topics and tasks. The associations' progress in building their capacities was assessed and tracked to help determine what further trainings they would need. Results of the assessment indicated the increased capacity of the community associations in the five components from the baseline (Fig. 13.3). The spider graph shows the mean values that demonstrate the improvements of the seven IPLC associations after the capacity-building trainings. Increased capacities were demonstrated for all five areas of assessment, each representing enhanced resilience to deal with current and future challenges. The widest range of change in capacity was found in the area of human resources, while financial management had the tightest array. Each of the IPLCs had a unique capacity growth profile (Fig. 13.3), representing the need for individualised planning for future activities through facilitated action research cycles (Watts and Pajaro 2014) to improve resilience.

The effectiveness of the capacity-building trainings in strengthening the community associations in the five components is further supported by the increased ratings generated by the Wilcoxon Signed Rank Test (Table 13.3). The positive ranks indicate a general increase in rating from the baseline for each of the components, although there were tied scores in some components of the assessment, i.e. they remained the same. The high absolute values of the Z scores resulted in low P-values, which describe the probability of the changes being random (Table 13.3). These results indicate that the capacities of organisations after training and capacity building were significantly higher than before the training and that community resilience was enhanced.

Table 13.3 Statistical summary for the institutional assessment using Wilcoxon Signed Rank Test

Community associations	Negative ranks	Positive ranks	Ties	Z score and P-value	
Kalipunan ng Malilit na Mangingisda sa Dagat ng Baler (KALMADABA)	0	42	11	−5.828	<0.001
Samahan ng Maliliit na Mangingisda sa Brgy. Matawe (Cabog)	0	41	12	−5.763	<0.001
DiCaDi Agta Group Tribal Association (DAGTA)	0	37	16	−5.684	<0.001
Samahan ng Katutubong Nangangalaga sa Lupaing Ninuno ng Dibut (Dibut SKNLND)	0	51	2	−6.466	<0.001
Dimasalang Egongot Tribe Association Inc. (DETAI)	0	50	3	−6.353	<0.001
Samahan ng Nagkakaisang Katutubong Alta ng Diteki (Diteki SNKAD)	0	28	25	−4.963	<0.001
Provincial Federation of Egongot Tribe of Aurora (PFETA)	0	39	14	−5.988	<0.001

Source: Authors, based on this study

The IPLCs allowed the communities through the association leaders to participate in several capacity-building activities and in turn facilitated the members to act together and benefit from the opportunities. For example, being able to manage finances in a transparent way and report on them to the scrutiny of the members was a new and useful experience for communities. All except one of the seven partner IPLCs had never opened a bank account. Submitting documents on their work and financial plans and reports, even if written in the Filipino dialect, was a struggle for IPLC project partners, but a huge accomplishment contributing to future efforts on building resilience. The said documents were required for the IPLCs' budget allocations to be released, according to the agreed terms. Towards the end of project implementation, agreements and plans on how their organisations could continue what they started became critical points of discussion. Of the IPOs with stronger leaders, Dibut's SKNLND was able to use members' contributions as capital and with further guidance, set up a store which sold basic commodities to its members at a very affordable cost. Another organisation, DETAI, reorganised as Dimasalang Egongot Tribe Farmers and Weavers Association, Inc. (DETFAWAI) in 2019 and registered with the SEC. Subsequently, having built a track record for successful project management, the organisation submitted proposals and was approved for funding amounting to 2.7 million PHP (US$47,000) from the Forest Foundation of the Philippines, as well as one million PHP from the Department of Environment and Natural Resources (DENR) to implement projects on biocultural heritage, protection and conservation of the ancestral domain, and to support its weaving industry.

Through the various capacity building activities implemented, the IPLC associations strengthened their legal and institutional status, thereby establishing track records with which to access future funding opportunities. Legal status allowed them to establish internal rules and policies that are relevant and connected to their culture and values. The Indigenous communities ensured that clan representation was secured during the selection of leaders with the approval of the tribal council and elders. For the fisherfolk, selection of leaders depended on active participation and knowledge of fisheries resources, which are important for leaders to establish authority and respect and to effectively lead collective action towards conservation. Good governance starts with selecting community leaders who go through the proper selection process with culture and traditions as the foundation. Recognising these customary institutions is said to further promote the understanding of restoration efforts and therefore increase local participation (e.g. Wehi and Lord 2017; de Koning et al. 2011). This process is also important for an association's accountability and conflict management.

The capacity building conducted over the span of 2 years (2016–2018) also provided the IPLCs with mechanisms to improve their internal systems in administration, human resources, and financial management. While some aspects of the institutional assessment values stayed the same, showing the need for further training and monitoring, the associations nevertheless grew significantly from their baselines. This growth has further strengthened the foundations for their rights and participation in local decision-making. They were recognised by external institutions, both from the government and private sectors, for future partnerships and collaborations. The Egongot IPLC through DETAI and PFETA were able to secure other funding to continue their conservation and restoration work within their ancestral domains. In addition, KALMADABA, DETAI, SKNLND, and SNKAD were able to partner with the national agency, DENR, for scaling up efforts in restoration through the National Greening Project. These capacity development activities potentially increased the resilience of the IPLCs by preparing them to proactively respond to future challenges, such as those emerging from climate change.

13.3.2 Actualising Local Synergies

Ecosystem Restoration and Protection Efforts
The trainings undertaken by the IPLCs on biodiversity conservation, nursery establishment, ecosystem restoration, and natural resources management eventually contributed to: (1) reforestation of 140 ha of denuded ancestral lands to support watersheds (Fig. 13.4a, b), where at least eight threatened endemic forest species were outplanted; (2) establishment of a 30-ha seagrass-mangrove protected area and 20-ha coastal greenbelt (Fig. 13.4c); and (3) updating the marine protected area management plan for the ancestral waters of San Luis (Fig. 13.4d).

These community-based restoration activities led by the IPLCs were catalysed by partnership with the provincial Technical Working Group (TWG), whose multi-agency members include representatives of an NGO (Daluhay), municipal and provincial local governments, national agencies (DENR and BFAR), and the academe (Aurora State College of Technology), with funding through SGP5. In Baler

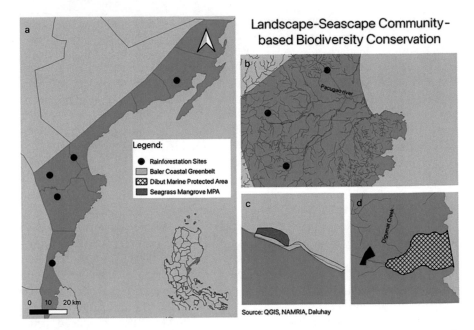

Fig. 13.4 Map of landscape-seascape restoration efforts of the IPLC (Source: Daluhay unpub. report 2018)

municipality, the participatory coastal resource assessment (PCRA) of the seagrass and mangrove area led by the KALMADABA fisherfolk association became the basis for the establishment and later adoption of a municipal ordinance designating mangrove-seagrass conservation management zones, including the protection of *Kandelia candel,* a rare mangrove species. Follow-up PCRA monitoring in 2018 indicated some improvement based on the recorded presence of the commercially valuable chiton mollusc, which was previously absent in 2016.

There were also restoration efforts led by the local government unit (LGU) through the Municipal Agriculture Office but still catalysed by the provincial TWG. In San Luis municipality, the LGU provided the funds and invited the IPLC to join in the participatory resource assessments of coastal and agricultural ecosystems. The proposed coastal management zones and prescriptions were presented to the community for recommendations and endorsements. These were then incorporated in the draft management plan and became the basis for a municipal ordinance. Efforts to create a model ridge-to-reef approach to biodiversity conservation led to the establishment of marine protected areas as OECMs. The establishment of an agricultural protected area (APA) was also initiated by farmers practicing a "pure" or mixed method of organic and synthetic farming. APA zones have been identified and were included in the proposed ordinance presented to the Municipal Agriculture Office for endorsement and to the Environment Committee chair of the Sangguniang Bayan (municipal elected council).

These efforts led to the formulation of a management plan for a network of locally managed marine protected areas covering 418 ha of coral reef (inclusive of the ancestral waters mentioned in Fig. 13.4d), 10 ha of marine turtle nesting area on the beach, and 6.56 ha of seagrass conservation zone. Another 18 ha of agricultural land, promoting biodiversity-friendly technologies, was proposed for establishment. The results of a fish catch monitoring survey in 2017 indicated that catch appears to have increased from 26 kg in 2015 to 32 kg on the average ($n = 224$), suggesting the positive effect of the efforts by the municipality and its partners to augment declining fish catch and potentially increase the income of San Luis fisherfolk. The farmers involved practiced conventional farming, but were also partly advocating the use of organic fertilisers and biodiversity-friendly pesticides. Of note, they claimed that it was not commercially profitable to fully practice the more sustainable farming methods. The IPLCs' local stewardship, resource use allocations, and protection from threats within the framework of their formulated resource management plans, as well as their organisational work and financial plans, also contributed to effective conservation. The connectivity of ridge-to-reef ecosystems and among the IPLCs added value by bringing a wider coverage of restoration and enforcement. Restoration in the watersheds (Fig. 13.4a, b) greatly affected the health of freshwater and marine ecosystems (Fig. 13.4c, d). Efforts to establish marine protected areas and coastal greenbelts also provided protection to upland communities and agricultural areas. The complementation of conservation efforts is crucial for community resilience in areas prone to typhoons and calamities, such as Aurora Province. Social inputs (traditions, values, and culture) are complemented by scientific inputs in biodiversity assessment and monitoring as part of the IPLC capacity building in resource management, strengthening the socio-ecological ecosystems and ethnoecology (Watts et al. 2021). Conservation and protection of the intertidal zones, where women and children usually gather seafood, provides a model for OECMs on optimising food security, improving water quality, and improving maternal nutrition (Alejos et al. 2021).

Food Security and Sustainable Livelihood Development
There are other enabling mechanisms arising from capacity and restoration efforts that are important for both social and ecological outcomes, food security, and livelihood development. Together these efforts critically enhance socio-ecological resilience, encompassing both ecocentric and anthropocentric parameters, steps forward in actualising an Ecohealth balance between the ecosystem approach to health and the health approach to ecosystem (Watts et al. 2015). In San Luis, the LGU has acknowledged that the strong technical support provided by the project allowed it to win a major prize in the *Malinis at Masaganang Karagatan* competition, sponsored by the Bureau of Fisheries and Aquatic Resources, a national agency. The prize, in the form of livelihood funds, was also based on the results of the socio-economic and biophysical profiling conducted through SGP5. It became apparent that deep sea fishing was being monopolised by big commercial fishing operators who were mostly based outside of San Luis, an illegal encroachment on municipal waters. At least five of the nine commercial vessels had deployed fish

aggregating devices (FADs) within the 15 km from the shoreline boundary of the municipal waters, where supposedly no commercial fishing should be allowed. Moreover, each fish aggregating device was placed at an average of only 4 km apart, thus altering the entire coastal ecosystem. This is a cause of concern throughout the municipal waters of Aurora Province, as reported by other coastal municipalities. This concern also ties in well with the initiative that collaborates with the Marine Resource Environment Foundation (MERF), also a SGP5 grantee, which aims to establish a North Philippine Sea (NPS) bioregional network, encompassing an assemblage of biodiversity that support fisheries critical to the food-security of more than ten million Filipinos, as well as undetermined local livelihood revenues. The NPS is highly dependent on pelagic or offshore fishery and the MERF project has looked into nearshore and offshore linkages that may require more policy research. The mayor of San Luis subsequently sent letters to the concerned commercial fishing vessel operators, warning them of their violation. Also, to address this inequitable access to deep sea fishery resources, the artisanal fishers have decided to form a cooperative and venture into sustainable fishing as a social (group initiated) and biodiversity-friendly enterprise.

Another of the IPLCs, a community association partner (i.e. PFETA— Table 13.3), has initiated the construction of a demonstration tribal village to showcase the traditional Egongot house and function hall. The function hall is also planned to serve as a mini museum that will house traditional tools/implements, instruments, and apparel. A brochure has been prepared to document different techniques in building the traditional Egongot house and the cultural meaning of the structures, as well as weaving and crafting traditional implements, instruments, outfits, and accessories. The use of herbs to treat some ailments is also a known traditional practice. Herbal products such as *lagundi* prepared as cough syrup and malunggay leaves as vitamins were produced as a potential income-generating venture that demonstrated traditional medicines used by ancestors. The documentation of Indigenous knowledge, systems, and practices, and the construction of a heritage village showcasing the unique culture of the Egongots, has inspired other Indigenous community leaders from other IPLCs engaged in the project to replicate the same concept in their areas as a way of preserving their heritage. In further examples of inter-community synergies, chieftains of the Egongots in other settlements (e.g. Dipaculao and Dimasalang) are considering similar demonstration villages with museums in their communities as part of their future plans. In this instance, cultural and biodiversity heritage played an important role not only in the process of co-designing restorations efforts and management plans but also for food security and livelihoods.

Networking and Linkages

Scaling up of SEPLS initiatives is a crucial strategy for a sustainable planet, requiring expansion of contacts outside the jurisdictions of the IPLCs. However, linkages within the same locality of the IPLCs should continue to be built and strengthened. The project advanced the landscape and seascape approach as it expanded its reach to cover the whole stretch of Aurora Province and beyond, embracing the other provinces along the North Philippine Sea through MERF, another SGP5 grantee. In Aurora Province, Daluhay has maintained close linkages with the IPLCs through their LGUs, particularly the Municipal Environment and Natural Resources Offices and/or the Agricultural Offices. Decadal Daluhay facilitation of provincial-level collaboration led to the establishment and active involvement of the provincial Technical Working Group (TWG), engaging academe, the provincial local government (Environment and Natural Resources and Provincial Agriculture Office), and national agencies including the Bureau of Fisheries and Aquatic Resources (BFAR) and the Department of Environment Natural Resources (DENR). The TWG played a crucial role in strengthening the IPLCs as it provides collaborative technical support. This has allowed LGUs from different municipalities to have easier access to technical expertise, which is required when establishing protected areas and OECM initiatives. Whenever possible, Daluhay also facilitated inter-community communication and event planning to promote exchange between the IPLCs. This approach enhanced resilience through shared leadership. Some communities were stronger in specific capacity building activities and were able to facilitate a broader sense of community around shared socio-ecological goals. Recognising that future action research cycles would first require establishing the next step priorities for each IPLC, further multi-IPLC projects have great potential to optimise both available resources and resilience enhancement.

Partnership with the provincial TWG and funding support from the SGP5 assisted in determining long-term goals and achieving short-term goals. The technical and financial support provided the necessary incentives for the implementation of restoration efforts and establishment of networking strategies by the seven IPLCs. Local policy frameworks are crucial for maintaining restoration efforts and further strengthening the IPLCs, as well as inclusion of conservation efforts in the economy through payment for ecosystem goods and services. The efforts of the IPLCs and TWG have led to institutionalised local policies: the Baler seagrass-mangrove protected area and the San Luis network of marine protected areas. In addition, a draft provincial resolution in support of the IPLC network that includes recognition and support for Indigenous Community Conserved Areas (ICCAs) has been submitted to a provincial councillor for sponsorship.

The SGP5 project also provided opportunities for community associations to closely collaborate with other SGP5 grantees outside of Aurora Province belonging to the Lower Sierra Madre Hub. These included the Bagong Lumad Artists Foundation, Inc. (BLAFI) and MERF. BLAFI worked with IPLC partners using social artistry as a tool to document local knowledge, systems, and practices related to biodiversity conservation. This documentation has been compiled into a collection of original song compositions in an album called *Awit ng Aurora* (see https://sites.google.com/a/blafi.org/main/awit-ng-aurora-album). Meanwhile, work with MERF allowed for expansion of the network to include the ten provinces around the North Philippine Sea and taking the lead in advancing the socio-economic concerns of stakeholders around the newly claimed offshore Philippine territory, the Benham Rise.

The efforts previously outlined in Sect. 13.3.1 on institutionalisation and capacitating IPLC, as well as Sect. 13.3.2, actualising local synergies, can be used to assess community resilience within the SEPLS framework. Due to the limited nature and scope of the activities implemented within the SGP5 project cited in this case study, we were able to assess only 17 of the 20 indicators described in the SEPLS resilience toolkit (UNU-IAS et al. 2014). The three indicators that were left out have been acknowledged as significant research gaps. These were: (1) diversity of local food system, (2) maintenance and use of local crop varieties and animal breeds, and (3) human health and environmental conditions. At the time of writing this chapter, we are filling these gaps by working with the IPLCs on a maternal ecohealth (Alejos et al. 2021) project considering sustainable community-based food systems, supported by the Satoyama Development Mechanism and the Aurora Provincial Health Office. A systematic resilience assessment using the UNU-IAS et al. (2014) toolkit was not considered as the two initiatives evolved simultaneously. For integrated planning and policy development, it would be appropriate to conduct focus group discussions with IPLCs when the opportunity arises.

13.4 Conclusions

Complex pathways and dynamic interactions persist between human culture and traditions and diverse ecosystems as experienced with the seven IPLCs in Aurora Province. To ensure sustainable ecosystem goods and services from these diverse ecosystems, a balance between production, conservation, and management should be maintained across the landscapes and seascapes, scaling up more localised ecohealth balance (Watts et al. 2015). The SGP5 implementation experience in Aurora Province demonstrated the significance of IPLCs as local institutions that can advance effective stewardship efforts in restoration of SEPLS through several traditional activities. These include sustainable self-regulation of resource use; habitat restoration; assertion of territorial rights; resistance to external extractive pressures; and the maintenance of stewardship practices under environmental, economic, and political change.

The various knowledge, skills, and experiences imbibed by IPLC project partners, Daluhay, the TWG, and the leaders of organised associations led to advancements in local management, planning, and resilient governance. These strategies are being used as an integral component of a nationalised effort to certify and institutionalise socio-ecological (Coastal Resource Management) skills (Pajaro et al. 2022); potentially leading to an international accord. In part, these efforts are being continued through an ongoing project with Saskatchewan Polytechnic University in Canada, working on micro-certification with Indigenous and artisanal communities in both countries. International competency accords have been instrumental in advancing institutionalisation within other fields, for example Engineering (Accord 2013). A similar action research result of establishing an accord for socio-ecological production system skills competency could help to expand global resilience. It is expected that monitoring will demonstrate increased social resilience with regard to planning and implementation based upon financial and human resource governance and administration (Fig. 13.3), leading to enhanced ecological resilience from restoration activities. The harmonised approach to socio-ecological resilience enhancement also served to break down barriers between communities and other governance agencies. This was a direct result of project synergies across communities and facilitators, such as those engaged in the province-wide TWG. The development of contacts by key community leaders supported collaborative efforts and was enhanced by participation in national-level conferences with SEPLS breakout sessions, such as the Lower Sierra Madre Hub, providing networking opportunities and the creation of broader more inclusive points of view connected to a national and international common framework. Reviewing and characterising project status, learnings, and challenges in the larger inter-community group helped in the organisation of consolidated actions based upon consensus. Further, these interactions and inter-community synergies met the goals of expanding ecosystem outcomes across jurisdictions, while maintaining the local integrity of each community group.

This case study demonstrates the key finding that embedding empowered local institutions and enabling mechanisms in conservation governance can lead to positive conservation outcomes through harmonised agendas and mandates. Likewise, the case study indicated that consideration and understanding of ILK, values, and political structure, and acknowledgement of their potential contributions to current ecosystem management regimens, are crucial for effective stewardship. Decision-making and conflict management processes are greatly affected by the IPLCs' progressive leadership, social ties, and norms. These underscore the most important social factors supporting attainment of positive conservation outcomes. The relevance of these social inputs demonstrated in this study support the idea of incorporation in national and regional policies through integration across scales of governance, local and national development plans. Formal recognition of IPLCs and incorporating ILK, strengthens IPLCs' legitimacy and enhances participation from community members, thus providing potential for more equitable and effective conservation. Local dependency on natural resources also puts them on the frontline in recommending appropriate SEPLS conservation and management measures. Government agencies, NGOs, and private organisations clearly still have individual roles to play in conservation, yet those roles may be constructively reoriented towards facilitation, supporting local capacity, and as a catalyst for scaling up of governance, while making efforts to understand local institutions cultural dynamics. These approaches can present cost-effective models of governance that deliver multifold benefits, supporting nature as well as social systems, equity, and accountability.

The capacity and subsequently the resilience of the seven IPLCs was expanded, resulting in more positive social and ecological outcomes through a harmonised management regimen. The project also provided a foundation for longer-term contributions to conservation and ongoing support for community resilience. In the Philippines context, this approach to broad province-based communication and cooperation towards biodiversity conservation and restoration with IPLCs can hopefully serve as a model and have a synergistic effect upon other biodiversity-rich provinces bordering the Sierra Madre Mountain Range, the North Philippine Sea, and beyond.

Acknowledgements The results of the case study discussed herein were made possible by the dedication and enthusiasm of the Daluhay team, partner Indigenous Peoples and Local Communities, critically, the municipal local government offices, based upon their Philippine mandate, as well as the Provincial Government of Aurora and national agencies including the Department of Environment and Natural Resources, the Bureau of Fisheries and Aquatic Resources, and the National Commission on Indigenous Peoples. The strategic project was awarded to Daluhay through the Global Environment Facility's 5th Operational Phase of the Small Grants Programme with the United Nations Development Programme as the implementing agency.

References

Accord Washington. Sydney Accord, and Dublin Accord (2013) Graduate attributes and professional competencies, international engineering alliance

Alejos SF, Pajaro MG, Raquino MR, Stuart A, Watts P (2021) Agricultural biodiversity and coastal food systems: a socio-ecological and trans-ecosystem case study in Aurora Province, Philippines. Asian J Agric Dev 18(2):12 pp

Comberti C, Thornton TF, de Echeverria VW, Patterson T (2015) Ecosystem services or services to ecosystems? Valuing cultivation and reciprocal relationships between humans and ecosystems. Glob Environ Chang 34:247–262

Cuerrier A, Turner NJ, Gomes TC, Garibaldi A, Downing A (2015) Cultural keystone places: conservation and restoration in cultural landscapes. J Ethnobiol 35:427–448

Daluhay Daloy ng Buhay, Inc (2018) Synergistic and ecocentric capacitation of Sierra Madre's Indigenous and Artisanal Communities. Unpublished Terminal Report, Submitted to UNDP-GEF SGP5

Dawson NM, Coolsaet B, Sterling EJ, Loveridge R, Gross-Camp ND (2021) The role of Indigenous peoples and local communities in effective and equitable conservation. Ecol Soc 26(3):19

de Koning F, Aguiñaga M, Bravo M, Chiu M, Lascano M, Lozada T (2011) Bridging the gap between forest conservation and poverty alleviation: the Ecuadorian Socio Bosque program. Environ Sci Pol 14:531–542

Ferraro PJ, Hanauer MM (2015) Through what mechanisms do protected areas affect environmental and social outcomes? Philos Trans R Soc B Biol Sci 370(1681):267

Garnett ST, Burgess ND, Fa JE, Fernandez-Llamazares A, Moln Z (2018) A spatial overview of the global importance of Indigenous lands for conservation. Nat Sustain 1(7):369–374

Golden CD, Allison EH, Cheung WW, Dey MM, Halpern BS (2016) Nutrition: fall in fish catch threatens human health. Nature 534:317–320

Katusiime J, Schutt B (2020) Linking land tenure and integrated watershed management—a review. Sustainability 2020(12):1667

Ong PS, Afuang LE, Rosell-Ambal RG (eds) (2002) Philippine biodiversity conservation priorities: a second iteration of the national biodiversity strategy and action plan. Department of Environment and Natural Resources-Protected Areas and Wildlife Bureau, Conservation International Philippines, Biodiversity Conservation Program-University of the Philippines Center for Integrative and Development Studies, Foundation for the Philippine Environment, Quezon City, Philippines

Pajaro M, Watts P, Ampa J (2013) The northern Philippine sea: a bioregional development communication strategy. Soc Sci Diliman 9(2):49–72. ISSN 1655-1524 Print / ISSN 2012-0796 Online

Pajaro M, Raquino M, Watts P (2022) Professionalizing community-based coastal resource management (CRM) services. J Ecosyst Sci Eco-Gov 4(Special Issue PAMS16):12–22. PRINT ISSN 2704 4394 I ONLINE ISSN 2782 8522

Pecl GT, Araújo MB, Bell JD, Blanchard J, Bonebrake TC (2017) Biodiversity redistribution under climate change: impacts on ecosystems and human well-being. Science 355:6322

Sangha KK, Russell-Smith J (2017) Towards an Indigenous ecosystem services valuation framework: a north Australian example. Conserv Soc 15:255–269

Strassburg BN, Iribarrem A, Beyer HL, Cordeiro CL, Crouzeilles R (2020) Global priority areas for ecosystem restoration. Nature 586:724–729

Taheri SM, Hesamian G (2013) A generalization of the Wilcoxon signed-rank test and its applications. Stat Papers 54:457–470

Tobón W, Urquiza-Haas T, Koleff P, Schröter M, Ortega-Álvarez R (2017) Restoration planning to guide Aichi targets in a megadiverse country. Conserv Biol 31:1086–1097

UNDRIP (United Nations Declaration on the Rights of Indigenous Peoples) (2013) A manual for National Human Rights Institutions. Asia Pacific Forum of National Human Rights and the Office of the United Nations High Commissioner for Human Rights, Switzerland

UNU-IAS, Bioversity International, IGES, UNDP (2014) Toolkit for the indicators of resilience in socio-ecological production landscapes and seascapes (SEPLS). https://collections.unu.edu/eserv/UNU:5435/Toolkit_for_the_Indicators_of_Resilience.pdf

Uprety Y, Asselin H, Bergeron Y, Doyon F, Boucher JF (2012) Contribution of traditional knowledge to ecological restoration: practices and applications. Ecoscience 19:225–237

Watts PD, Pajaro MG (2014) Collaborative Philippine-Canadian action cycles for strategic international coastal ecohealth. Can J Action Res 15:3–21

Watts P, Custer B, Yi ZF, Ontiri E, Pajaro M (2015) A Yin-Yang approach to education policy regarding health and the environment: early-careerists' image of the future and priority programmes. Nat Res Forum 39(3–4):202–213

Watts PD, Pajaro MG, Raquino M, Anabieza J (2021) Philippine fisherfolk: sustainable community development action research and reflexive education. Local Dev Soc. https://doi.org/10.1080/26883597.2021.1952847

Wehi PM, Lord JM (2017) Importance of including cultural practices in ecological restoration. Conserv Biol 31:1109–1118

The opinions expressed in this chapter are those of the author(s) and do not necessarily reflect the views of UNU-IAS, its Board of Directors, or the countries they represent.

Open Access This chapter is licenced under the terms of the Creative Commons Attribution-NonCommercial-ShareAlike 3.0 IGO licence (http://creativecommons.org/licenses/by-nc-sa/3.0/igo/), which permits any noncommercial use, sharing, adaptation, distribution and reproduction in any medium or format, as long as you give appropriate credit to UNU-IAS, provide a link to the Creative Commons licence and indicate if changes were made. If you remix, transform, or build upon this book or a part thereof, you must distribute your contributions under the same licence as the original. The use of the UNU-IAS name and logo, shall be subject to a separate written licence agreement between UNU-IAS and the user and is not authorised as part of this CC BY-NC-SA 3.0 IGO licence. Note that the link provided above includes additional terms and conditions of the licence.

The images or other third party material in this chapter are included in the chapter's Creative Commons licence, unless indicated otherwise in a credit line to the material. If material is not included in the chapter's Creative Commons licence and your intended use is not permitted by statutory regulation or exceeds the permitted use, you will need to obtain permission directly from the copyright holder.

Chapter 14
Synthesis: Ecosystem Restoration in the Context of Socio-Ecological Production Landscapes and Seascapes (SEPLS)

Maiko Nishi, Suneetha M. Subramanian, and Alebel Melaku

14.1 Concept: Can Landscape Approaches Underpin Ecosystem Restoration?

Landscape approaches are characterised by an explicit recognition of social-ecological systems in different contexts, resulting in interventions that concurrently address anthropogenic issues of concern to people, as well as biodiversity decline, and ecosystem degradation. Furthermore, a social-ecological paradigm can also be used to address concerns across multiple stakeholders to help ensure that conservation and sustainable use of natural resources are performed in an equitable manner.

Optimally, ecosystem restoration is best supported by a multipronged and transdisciplinary approach to addressing underlying natural and anthropogenic drivers of degradation. In this context, landscape approaches could potentially play a pivotal

Contributing authors to this chapter include Archana Bhatt, Nancy Chege, Dhanya Sreenivasan Chemboli, Jyun-Long Chen, Camila I. Donatti, Devon Dublin, Godwin Evenyo Dzekoto, Kizito Echiru, Siddharth Edake, Ernest Ngulefack Forghab, Alexandros Gasparatos, Ade Bagja Hidayat, Kang Hsu, Paulina G. Karimova, Tom Kemboi Kiptenai, Chunpei Liao, Yufen Chuang, Guanqi Li, Hwan-ok Ma, Yoji Natori, Jacqueline Sapoama Mbawine, Anil Kumar Nadesapanicker, Louis Nkembi, Njukeng Jetro Nkengafac, Josephat Mukele Nyongesa, Samuel Ojelel, Raymond Owusu-Achiaw, Marivic Gasamo Pajaro, Dambar Pun, Vipindas P., Mark Edison R. Raquino, Pia Sethi, Xin Song, Jie Su, Aashish Tiwari, Tamara Tschentscher, Yaw Osei-Owusu, Paul Watts and Chemuku Wekesa.

M. Nishi (✉) · S. M. Subramanian · A. Melaku
United Nations University Institute for the Advanced Study of Sustainability (UNU-IAS), Tokyo, Japan
e-mail: nishi@unu.edu

© The Author(s) 2023
M. Nishi, S. M. Subramanian (eds.), *Ecosystem Restoration through Managing Socio-Ecological Production Landscapes and Seascapes (SEPLS)*, Satoyama Initiative Thematic Review, https://doi.org/10.1007/978-981-99-1292-6_14

role. Some of the most relevant features of these approaches that enable effective and successful restoration include:

People and Their Practices Typically, SEPLS are sites of diverse resources, mosaic ecosystems, and multiple stakeholders who relate to the landscape or seascape in various ways. To the local communities, SEPLS provide a means of livelihood, enable attainment of basic material needs (e.g. food, timber, and water), and are linked to health and well-being (e.g. physical and mental health, sense of identity or belonging, and other cultural values). To those engaged in commercial activities, they provide opportunities for economic gain from both within and outside the landscape and/or seascape through trade in natural resources-related products and services. For local and national administrators, there is an ongoing need to address issues of sustainable use and the reconciliation of conservation, restoration, and livelihood needs, particularly within large-scale development activities that may include land and sea use change.

Balancing ecocentric and anthropocentric drivers is considered to be a priority for the long-term health of ecosystems across several development settings, as well as for human health and well-being (Watts et al. 2015). Optimising SEPLS management for restoration requires a common understanding, across different stakeholders, of historical land and sea uses, the competing landscape needs of the users, and alignment with relevant policy goals. It follows therefore that social-ecological diversity would form the basis to identify and implement different types of solutions (cutting across ecological, economic, and social parameters) that work in various contexts—e.g. peace parks in Nepal (Chap. 4), the system of producing several varieties of rice staple in India (Chap. 8), Ridge-to-Reef food systems and marine bioregional engagement in the Philippines (Chap. 13), and shifting cultivation systems under changing socio-economic circumstances in India and Thailand (Chap. 7).

Knowledge Assets Given the diversity of actors and resources, SEPLS sites are often rich in Indigenous and Local Knowledge (ILK) that has been held, innovated, practiced, and developed over time by Indigenous Peoples and Local Communities (IPLCs) and further, scientific and expert knowledge. Appropriate integration of these different systems of knowledge is often practiced towards necessary solutions in the landscape and/or seascape—e.g. participatory tree nursery species selection and reforestation strategies in the Philippines (Chap. 13), rangeland restoration in Chyulu, Kenya (Chap. 3), sacred groves reservation in Ghana (Chap. 2), Indigenous and Community Conserved Areas (ICCAs) in South China (Chap. 12), and the use of traditional knowledge on tree species to support ecosystem restoration in fallow lands in India and Thailand (Chap. 7).

Synergistic Governance Approaches The wide range of actors involved in the use and management of a landscape and/or seascape (and the mosaic of ecosystems therein) necessitates co-operative approaches to decision-making to ensure management activities are oriented towards promoting restoration and sustainable use. These involve ensuring cross-sectoral partnerships that allow for policy coherence and sustaining multifunctionality of the landscape or seascape. They ensure a nested system of governance from international (wherever appropriate) to national to local and thereby are also inclusive of local cultural norms and traditional knowledge, while fostering collaborative management (Watts et al. 2021). Such approaches can be facilitated through multi-stakeholder platforms and harmonised objectives between different actors (e.g. Chaps. 10–13). They recognise the various rights of the communities, enabling capacity development and awareness raising of different stakeholders on the interdependence and interconnections between environmental health and human well-being—e.g. Community Resource Management Committees (CRMCs) in Ghana (Chap. 9), conservation of rice-based ecosystems in India (Chap. 8), Ecohealth approaches and forest management in the Philippines (Chap. 13), the Community Resource Management Area (CREMA) system in Ghana (Chap. 2), and traditional resource management practices in India and Thailand (Chap. 7).

Leveraging landscape approaches for ecosystem restoration that benefits biodiversity, ecosystems, and human well-being in SEPLS can be undertaken by paying attention to the following:

- Identifying and addressing drivers of degradation (e.g. production practices, invasive species, policies, and the political and economic causes of these drivers).
- Leveraging cross fertilisation of knowledge, including scientific, expert, and ILK (referred to as knowledge weaving) (Tengö et al. 2014).
- Identifying the potential trade-offs between biodiversity conservation/restoration and human needs (e.g. livelihood security, human–wildlife conflict, large-scale economic development, and other subsistence-related activities).
- Quantifying costs and losses (to household incomes, private sector revenue, and public infrastructure) from ecosystem degradation and related impacts, focusing on the costs of action vs. inaction.
- Linking restoration activities with priorities and needs of IPLCs.
- Broadening the scope of ecosystem restoration to include health (including that of people, animals, and environment), well-being, and economic development in consonance with the priorities of all relevant stakeholders who need to act to ensure social-ecological resilience.
- Taking stock of emerging issues—including climate change, new forms of production practices, pandemics, and ways to interlink disciplinary and sectoral interventions (e.g. nexus approaches viz. One Health, Ecohealth, food-water-energy, and disaster risk reduction) that synergise across multiple environmental and developmental goals.
- Supporting community-driven development of nature-based value chain enterprises.
- Fostering multi-stakeholder platforms to ensure multiple voices, including those of youth and women in communities and of the private sector and special interest groups (e.g. faith-based organisations), is considered to build consensus through their participation in policy and decision-making.

- Enhancing enabling factors such as raising public awareness, mobilising financial resources, developing capacities to tackle the complexity of issues, advancing adaptive co-management practices (inclusive of monitoring and evaluation), and striving to foster political will towards restoration actions.
- Identifying and responding to priorities of different stakeholders and determining timely and reflexive actions.

Embedding Biodiversity in Implementing Landscape Approaches to Ecosystem Restoration

While ecosystem restoration could be pursued through different pathways, a special interest at the SEPLS level is to make sure that activities aimed at ecosystem restoration are biodiversity friendly to ensure the sustenance of bio-cultural heritage and diversity and human well-being. Towards this, some pragmatic approaches are highlighted below:

- Explicitly identifying how restoration practices will promote biodiversity conservation and restoration, and the resultant trade-offs.
- Ensuring appropriate incentives to land managers and different stakeholders. These could range from those appealing to economic (e.g. innovative markets for various ecosystem functions such as water regulation, payment for ecosystem services (PES), and carbon funds), and cultural values (e.g. sense of place, cultural identity, educational value, and aesthetics), to those related to a better quality of life (e.g. disaster risk reduction, biodiversity-based livelihoods such as ecotourism, non-timber forest products (NTFPs) based livelihoods, and successional agroforestry).
- Promoting community conservation efforts and harnessing traditional knowledge that uses natural processes for ecosystem restoration (e.g. group farming, community-based farming committees, community conserved areas) that foster shared values.
- Customising landscape approaches to different ecosystems and social contexts to fit local conditions and resources [e.g. trans-ecosystem local food systems (Alejos et al. 2021)].
- Developing restoration plans that are based on comprehensive baseline assessments that clearly outline the degree of biodiversity richness and help establish local priority biodiversity goals.

The following sections highlight some methodologies, approaches, and strategies to explicate how this may be achieved.

14.2 Methodology

Given the large investments (e.g. finance, labour, and time) made in restoration activities, it is critical to assess their effectiveness. Such an assessment can also provide a basis for decision-making and management (Zhai et al. 2022) and improve best practices (Wortley et al. 2013).

Measuring the effectiveness of ecosystem restoration depends on well-defined and clear goals and objectives that specify the desired direction and magnitude of change (FAO et al. 2021). This also allows for clear communication of expected results, serves as a foundation for planning and implementation, and enables monitoring and evaluation and thus adaptive management. Nevertheless, ecosystem restoration frequently fails to achieve the expected goals due to the complexity and diversity of ecological, economic, social, cultural, and other elements associated with restoration (Hopfensperger et al. 2007), especially those relating to resolving trade-offs between multiple goals and objectives in a transparent and equitable manner (Villarreal-Rosas et al. 2021).

From the perspective of SEPLS, the measurement of restoration effectiveness necessitates holistic approaches that integrate the ecological, social, cultural, and economic dimensions of changes in the landscapes and seascapes. This section discusses how we can measure, evaluate, and monitor the effectiveness of SEPLS management to prevent, halt, and reverse degradation and achieve restoration objectives. Specifically, we discuss the benefits of measuring effectiveness, indicators to gauge and evaluate effects and outcomes, and tools for measuring the effectiveness of restoration through SEPLS management.

What Are the Benefits of Measuring Effectiveness (Both Tangible and Intangible) that Have Emerged from Ecosystem Restoration Through a SEPLS Lens?
Measuring the effectiveness of ecosystem restoration in the context of SEPLS management has numerous benefits ranging from improvement in providing goods and services for community livelihoods to better provision of intangible benefits such as knowledge transfer and conservation of ILK associated with ecosystem restoration and management. The following are some of the advantages of measuring the effectiveness of ecosystem restoration:

- *Transfer knowledge and share lessons learnt:* Knowledge transfer on SEPLS management is crucial to measuring ecosystem restoration by various stakeholders including future generations. Through inter-generational transfer, knowledge is handed down on how natural resources can be preserved and sustained. Practices and experiences of ecosystem restoration and management building on local knowledge can be used as lessons for future restoration projects and scaling up of such practices and innovations. Furthermore, they can be unpacked to the public, and other stakeholders interested in restoration can use the results and procedures to develop or improve their projects. Importantly, stakeholder participation during the span of the restoration project can foster knowledge acquisition, support capacity development, and enhance connectivity among stakeholders at the local, national, and global levels. This facilitates valid knowledge transfer and sharing for scaling up and out the restoration, as seen in the case studies herein (e.g. Chaps. 5, 10, and 13).

- *Facilitate adaptive management:* Because the process of restoration is normally lengthy, changes in conditions are almost unavoidable. New data and ideas could be incorporated into the planning and implementation of a project to ensure that the restoration objectives are achieved effectively despite the inevitable changes. This not only supports success in restoration, but also promotes adaptive management of SEPLS (e.g. Chap. 7).

- *Support contribution to human well-being and biodiversity:* Ecosystem health and human well-being are equivalently important within the SEPLS concept. Restoration activities have a vital role in recovering and sustaining biological and functional diversity, mitigating and adapting to climate change, and maintaining the livelihoods of local communities. In this regard, the evaluation of effectiveness can show, for instance, how much restoration helps to recover biodiversity and improve the well-being of local communities at the same time, and how much it enhances community resilience to climate change (e.g. carbon sequestration and storage, flood control, and erosion control). Furthermore, measuring the effectiveness of restoration through the valuation of ecosystem services allows for attaching values to certain ecosystem services that are not directly traded within a market or lacking in defined market prices (e.g. willingness to pay and hedonic pricing). As such, the evaluation helps explore ways to ensure that the communities and biodiversity can continuously and sustainably benefit from the restoration (e.g. Chaps. 10 and 13).

- *Influence policy frameworks:* Restoration projects that are well aligned with local needs and government priorities help project implementers effectively engage stakeholders. When restoration activities are the local priority, local communities support the restoration process and may be willing to pay for it. More importantly, they have a strong sense of ownership. This enables effective participation in the activities and thus meaningfully affects the restoration programme. The same holds if projects align with government priorities. Working with responsible government institutions for a shared goal of ecosystem restoration can help attain the restoration objectives effectively by mobilising financial and human resources on a common agenda. The restoration activities could also align with businesses that wish to offset their corporate social responsibilities through restoration. Therefore, promoting and replicating restoration practices and methods based on evaluation helps to better design laws, policies, and plans at the local, national, and global levels to prevent, halt, and reverse ecosystem degradation. Policymakers at any level can use the results as a benchmark to formulate and improve policies for meaningful action involving numerous tangible and intangible benefits from SEPLS management (e.g. Chaps. 7, 10, and 13).

- *Define the success of restoration:* Success or failure rests on how to manage trade-offs among multiple effects of restoration (including positive and negative impacts on ecosystem services, biodiversity, livelihoods, and human well-being) while incurring restoration costs. Examining the trade-offs in consideration of the associated costs can help decision makers to better understand a full range of actual impacts of restoration, guide the goal setting of future restoration activities, and effectively allocate limited resources. In addition, although we always envisage restoration efforts to be successful, it is vital to adequately report on the failures and how they were addressed. Recording failures and remediation experiences can serve as learning points for future restoration projects.

What Indicators or Measurements Determine and Evaluate the Effects and Outcomes of Ecosystem Restoration Through Managing SEPLS?

Evaluating restoration effectiveness is a complicated task, as evident in numerous debates over what constitutes effective restoration and how to best evaluate it (Wortley et al. 2013). It depends on the specific objectives of the restoration project, which may have its own set of indicators for monitoring and evaluation. These indicators are essential for measuring effectiveness and play a crucial part in ecosystem restoration and management. Every effort needs to be made to ensure indicators or measurements are aligned with overall goals or objectives. For instance, restoration activities intended to mitigate the effects of one or more ecosystem challenges must at the very least consider how the success in this regard will be measured (Schultz et al. 2012). Although each restoration project has specific objectives and associated indicators, the following may be used as reference points to evaluate restoration outcomes:

- *Quality of life*: Improvements in the overall quality of life of those who live in a landscape or seascape (e.g. impacts on food and water security, livelihoods, health, and human well-being) can be used as an indicator to gauge the impact of restoration on the ecological, economic, and social dimensions of life in the community (e.g. Chaps. 6 and 13).
- *Natural capital*: Positive impacts of restoration on natural capital [i.e. stock of natural assets that includes soil, air, water, and living things (NCC 2013)] can be measured as improvements, like those in biodiversity and various natural systems of the Earth. This may manifest in terms of ecological functioning (e.g. nutrient cycling, erosion control, and carbon sequestration), land productivity, vegetation cover, natural regeneration, invasive species control, biodiversity richness, and air and water quality (e.g. Chaps. 2, 5, 9, 10, and 13).
- *Youth and women involvement and capacity development*: Creating enabling conditions for active participation of youth and women in restoration activities helps to transfer ecological knowledge to multiple actors thus facilitating the sustainable use and management of landscape and seascape resources. The involvement of youth also strengthens the long-term perspectives on ecosystem restoration activities. The effect of restoration on youth and women through capacity development, employment, and income generation can be considered an indicator (e.g. Chaps. 2, 4, and 13).
- *Community empowerment and participation*: Progress or achievements in active and meaningful participation of local communities in the restoration process from

the planning to the implementation stage can be considered as one of the measurements of restoration outcomes. This may include gender mainstreaming and capacity development of local communities in terms of knowledge, skills, social cohesion, and appreciation of nature's contributions to people through training, facilitation, and other community engagement approaches (e.g. Chaps. 2, 6, 8–11, and 13).

- *Sustainable Development Goals (SDGs)*: Ecosystem restoration through managing SEPLS can help achieve human well-being as captured by multiple relevant SDGs (FAO and UNEP 2022). For instance, from the human dimension, improved livelihoods (e.g. income and employment) support SDG 1 (No poverty), while improved food security and human well-being are linked explicitly to SDG 2 (Zero hunger) and SDG 3 (Good health and well-being), respectively. Other SDGs are related in terms of ecological aspects, including SDG 13 (Climate action), SDG 14 (Life below water), and SDG 15 (Life on land). The indicators of SDGs can be scaled down and up with careful consideration of potential cross-scale trade-offs to evaluate to what extent the restoration activities on the landscape or seascape scale help to contribute to the 2030 Agenda for Sustainable Development.
- *Climate resilience*: The contribution of SEPLS management to restoration can be evaluated in terms of resilience to climate-related risks, vulnerabilities, and impacts. For instance, restored vegetation cover could reduce the effects of flooding on livelihood assets and infrastructure, prevent soil erosion, and contribute to increased land productivity—e.g. participatory seed conservation and exchange programme in India (Chap. 8) and creation of a network of nurseries in the Philippines (Chap. 13).

What Are the Tools and Methods to Identify and Keep Track of the Effectiveness of SEPLS Management in Facilitating Ecosystem Restoration?

A comprehensive methodological framework would allow SEPLS managers to identify and address multiple needs and interests in the landscape or seascape and facilitate concerted efforts based on their findings for restoration and sustainability. Furthermore, if such a framework renders global consistency and comparability (e.g. universal monitoring standards), it would enable better communication with higher-level policy arenas and science-policy interfaces. The following are some of the tools, methods, and approaches that have helped to identify and keep track of the effectiveness of restoration in the context of SEPLS management. Some of the approaches and techniques were directly applied in the case studies presented in this volume (as reference chapters are indicated), while others were suggested by the practitioners and researchers who participated in the IPSI Case Study Workshop held in 2022.

- *Multi-stakeholder (participatory) approach*: This approach emphasises the active participation of community members, organisational affiliates, and researchers in all aspects of the restoration process (Israel et al. 1998). Participants contribute their expertise to improve their understanding of a given factor and incorporate the acquired knowledge into action to benefit the community involved. The people in the community should be part of the entire project cycle, including the planning stage, to build consensus and create a sense of ownership for sustainability, and the evaluation process, to have a shared valuation of the benefits gained from restoration (e.g. Chaps. 4, 6, and 8–13).

- *Indicators of Resilience in SEPLS*: This is a monitoring and evaluation tool that has proved effective in many SEPLS around the world (Dunbar et al. 2020). As a community-based participatory instrument used for assessing social-ecological resilience in SEPLS, the indicators [appropriately localised to SEPLS specifics in some instances (Karimova et al. 2022)] may also be applied on a regular and consistent basis to track the effectiveness of ecosystem restoration efforts in SEPLS over time and to understand the impact of the interventions (Dublin and Natori 2020).

- *Ecohealth paradigm*: This paradigm hinges on a transdisciplinary approach to addressing ecological, cultural, and socio-economic changes for sustaining and enhancing ecosystem and human health and well-being. This approach recognises ecosystems as another critical determinant of human health (Orlando et al. 2022) and has been suggested as a potential lens to balance the ecosystem approach to health and the health approach to ecosystems (Watts et al. 2015). It gives dual attention to the ecosystem and public health and facilitates bridging a gap between them that could be filled through ecosystem restoration (e.g. Chap. 13).

- *Action research*: To institutionalise SEPLS management inclusive of restoration activities, this approach provides a model for iterative designs that includes the steps of planning and assessment and evaluation that can be easily built into annual planning and funding regimens. Also considering international exchange and synergies (Watts and Pajaro 2014), it is recognised as a significant approach to sustainable development (Keahey 2021) (e.g. Chap. 13).

- *Biodiversity indexes and indicators*: Already available indexes for measuring changes in biodiversity can also be used to determine how effective the restoration is in improving biodiversity—e.g. Simpson index (Simpson 1949), Shannon Index (H′) (Supriatna 2018), Pielou's evenness index (Pielou 1969), and Menhinick's richness index (Menhinick 1964) (e.g. Chaps. 5, 7, and 13).

- *Information and Communications Technology (ICT) and emerging technology*: Currently available technological tools and devices can be used for collection, analysis, and reporting of the changes made by restoration activities [e.g. mobile applications, remote sensing, and Geographic Information Systems (GIS)]:

  – *Mobile-based technology*: Mobile phones with user-friendly applications can be used in field data collection for planning, monitoring, and evaluation. These technologies can also be used for real-time data sharing and analysis, which allows organisations, community members, and other stakeholders to monitor and evaluate data to improve planning and implementation (e.g. Chaps. 2, 4, 8, and 13).

  – *GIS and remote sensing*: GIS and remote sensing can be used to quantify the
    area restored through SEPLS management (Zhai et al. 2022). Remote sensing
    is a powerful tool for generating data for understanding and monitoring
    ecosystem restoration by obtaining information from electromagnetic radia-
    tion reflected or emitted from the Earth's surface and atmosphere. Researchers
    have used it for many years to assess vegetation phenology, land use and land
    cover change, species pattern and distribution, and to estimate vegetation
    biomass which is critical for understanding the impact of restoration on
    climate change and the carbon cycle (e.g. Chap. 3). GIS can be used to identify
    specific management boundaries for different land use priorities
    (e.g. Chap. 13). It can also play a critical role in putting certain concepts
    [e.g. landscape ecology (Turner et al. 2001) and reserve design (Peck 1998)]
    into practice, as well as in managing the geographic information that ecosys-
    tem restoration activities generate, which helps to assess progress and facilitate
    effective planning.

- *Capital assets framework*: This approach was adjusted from the sustainable
  livelihoods framework. It considers five kinds of capital assets (financial,
  human, natural, physical, and social) to understand livelihood outcomes and
  risk (Zhang et al. 2020). Financial capital includes credit and savings, pensions,
  and subsidies; human capital includes a wide range of human resources as well as
  social and personality traits such as education, skill, knowledge, health, and
  labour; natural capital includes natural resources such as forest resources, soil
  resources, and wild resources; physical capital includes the basic infrastructure
  and goods required to support livelihoods; and social capital includes social
  resources (Lax and Krug 2013; Zhang et al. 2020). The five capital assets can
  be evaluated using a variety of indicators, criteria, and principles. For the
  valuation and implementation of the restoration programme, beneficiaries can
  be asked to rate their perceptions of different indicators on a 1–5 Likert scale from
  1 (low) to 5 (high) (Ken et al. 2020).
- *Economics of ecosystem restoration*: In recent years, economic principles, tools,
  and instruments have become more extensively and comprehensively applied in
  restoration studies (Iftekhar et al. 2017). For instance, a new framework for
  ecological restoration includes the costs for project development, implementation, and maintenance, as well as opportunity costs and net income, while taking
  into account the economic value of recovered biodiversity and restored ecosystem
  services (Iftekhar and Polyakov 2021). The economics of ecosystem restoration
  thus provides information for better-evaluating restoration outcomes and helps
  decision makers to allocate scarce resources among alternative restoration projects. Also, The Economics of Ecosystem Restoration (TEER) was launched in
  2019 as a new multi-partner initiative in an attempt to estimate the net benefits per
  hectare per year of any restoration intervention in a given context.
- *The 5-Star Recovery System*: As a tool developed to track progress in ecosystem
  recovery (McDonald et al. 2016), the 5-Star Recovery System can assist managers, practitioners, and others in monitoring progress towards recovery goals

over time and assessing and ranking the degree of recovery of a site. This tool employs a 5-star scale representing a cumulative gradient of similarity to a reference ecosystem recovery state, ranging from very low to very high. An overall assessment can assign a restoration site to one of five recovery levels (1–5 stars).

14.3 How Can We Leverage Landscape Approaches in Restoring Ecosystems for More Sustainable Futures?

SEPLS management hinges on landscape approaches, by which diverse stakeholders are brought in to negotiate and collaborate for balancing multiple objectives in a given spatial area on land and/or sea. While place-based, multi-stakeholder approaches alone cannot address all challenges, they do provide opportunities for minimising trade-offs through collaboration, and for effectively restoring ecosystems to move towards more sustainable futures. The section below describes the challenges that actors regularly encounter in engaging in ecosystem restoration and managing SEPLS. It then discusses ways forward to tackle these challenges, develop synergistic strategies, and implement restoration activities for long-term SEPLS management.

14.3.1 Challenges in Ecosystem Restoration

Restoration efforts need to be planned and implemented in an integrated, participatory, and continuous manner to achieve restoration objectives (i.e. prevent, halt, and reverse land and sea degradation) and contribute to sustainable development. However, as shown in the previous chapters, actors operating at the landscape and/or seascape level face multiple challenges throughout the planning cycle to meet these goals. The challenges are mainly to (1) secure and otherwise raise capacities and resources for initiating the effort, (2) promote and facilitate the initiative in a concerted and coordinated way, and (3) sustain the effort and if needed, adapt it to changes.

Capacities and Resources
As a prerequisite, capacities and resources should be available and accessible for actors engaging in restoration activities. These capacities and resources range from intangible (e.g. knowledge, motivation, experience, trust, and innovation) to tangible ones (e.g. budget and finance, equipment, and labour). If these are already available or easily acquired, the actors can draw on them to plan and implement the activities autonomously. Yet, it is often the case that they are insufficient to fully grapple with restoration. Moreover, accessibility to the needed resources (e.g. time and technical

tools) varies among different stakeholders, whereby they are often least available to those who are most vulnerable to land or sea degradation. This makes the challenges in restoration more intractable.

Ecosystem restoration requires certain changes in the pathways through which people interact with nature. It is essential that local communities recognise the problems with their current trajectories and are motivated to embark on changes for better human–nature relationships (e.g. Edake et al. 2019). However, problems are not always salient and can be easily discounted or disregarded in everyday priorities. Furthermore, causal factors and relationships to the problems are often hard to comprehend. Indeed, SEPLS are dynamic, complex systems involving plenty of uncertainties, and thus require careful observation and continued monitoring for sustainable management. In particular, the management of seascapes, including underwater dimensions, epitomises the challenges in knowing, understanding, stewarding, and administrating such fluid, dynamic, and elusive systems that rapidly change in space and time without horizontal and vertical boundaries, while being influenced by various environmental and anthropogenic factors (e.g. climate change, currents, tides, temperature, microhabitats, and social and economic activities) (Maxwell et al. 2015). Moreover, in connection to the restoration level to be achieved, to what extent people can and should intervene in natural processes is highly debatable and controversial from a variety of standpoints encompassing ecological, social, cultural, and ethical dimensions (Filbee-Dexter and Smajdor 2019; Florentina-Cristina et al. 2017).

The ability to change the pathways for restoration primarily rests on local buy-in, acceptance, and agreement on new initiatives or interventions in human-nature interactions. By shifting the ways and means of living associated with nature, people in the local communities may lose the conventional forms and patterns of living on which they place value. Some case studies suggest that state policies often push their priorities and logic, while overlooking local needs, opportunity cost, implicit local contexts, and cultural nuances related to ecosystem restoration (e.g. Chap. 7). This could lead, for instance, to mismatches between given incentives and actual motivations for local participation, or a lack of local support for restoration initiatives. Failures in contextualising restoration in local settings may undermine the credibility of governmental interventions or erode trust in government authorities. This not only hinders synergies to be built among stakeholders, but also may result in inaction or even negative consequences for the well-being of local communities.

Restoration initiatives can be fruitfully informed by culture, ILK, and traditional practices that have historically ensured sustainable use and management of natural resources. Such culture and traditions can be carefully drawn on for restoration to make sure knowledge holders are respected and equitably rewarded, compensated, or benefited. To do so, Free, Prior, and Informed Consent (FPIC) is expected to be sought in the restoration projects. Interestingly, this FPIC process appears to be already built in for most cases of SEPLS management involving restoration activities where IPLCs are placed at the centre of the restoration initiatives and are taking the lead in improving their environmental and livelihood conditions (e.g. Chaps. 1 and 11–13).

Nevertheless, many communities today are faced with weakening or erosion of ILK, while experiencing rapid demographic, socio-economic, and ecological changes such as outmigration, urbanisation, technological development, and land abandonment (e.g. Chaps. 9 and 12). In this context, it is imperative not only to revive such culture and traditions, but also to bring in additional resources or develop new capacities to address emerging challenges and resolve unprecedented issues in managing SEPLS. This is important particularly when an intervention is introduced to local communities by external stakeholders or in partnership with multiple stakeholders. In some cases, the communities may be unprepared for new development (e.g. tourism, infrastructural development, and afforestation), disallowing them to concurrently engage in conservation practices (e.g. Chaps. 12 and 13). In other cases, externally funded projects could enable multi-stakeholder collaboration on restoration, but such synergistic actions may not happen unless some tangible resources (e.g. budget and human resources) are procured (e.g. Chaps. 10–13).

Coordination and Negotiation

To develop restoration strategies and facilitate their implementation, careful attention should be paid to trade-offs, including potential ones, manifested within and beyond a landscape or seascape. In the face of trade-offs within a landscape or seascape, reaching an agreement among stakeholders (e.g. local communities, regional and national governments, private sector, and NGOs) may be a challenge, given their diverse interests, priorities, and roles and responsibilities in restoration.

In this connection, rights and access to natural resources need to be deliberately examined, attending to political and economic power asymmetries between different stakeholders. For instance, the establishment of strictly protected areas may lead to positive environmental outcomes (e.g. biodiversity conservation). However, it could be detrimental to locals' rights to natural resource use by limiting their access to those resources on which they depend for their livelihoods (e.g. hunting restrictions imposed by authorities), or may raise short-term opportunity cost to locals who could otherwise generate income (e.g. limiting the use of mangrove forest for aquaculture). These trade-offs affecting the survival and well-being of local communities can happen particularly when communities are not involved in the decision-making process. Thus, it is crucial to empower and support the local community to participate in decision-making processes for restoration. Accordingly, such local participation and empowerment should be widely supported and ensured through implementation of principles of fair and equitable benefit-sharing. This would help to develop a self-sustaining system within a landscape and/or seascape (e.g. trans-ecosystem food systems in Chap. 13, and a peace park dealing with trade-offs between forestry and ecotourism in Chap. 4).

However, trade-offs can go beyond the readily recognisable temporal and spatial scale of a landscape or seascape, making the challenges rather tangled for coordination and negotiation. First, restoration effects can transcend a certain timeframe through interlinking with multiple social and ecological processes. For example, mangrove restoration aiming at both biodiversity conservation and ecotourism development within a seascape may result in the improvement of environmental quality only for a certain period of time; however, in the long run, it may instigate new development projects (e.g. infrastructural development and mining) alongside the flourishing ecotourism or may be influenced by continued or expanded industrial activities (e.g. Chaps. 9, 11, and 12). This may in turn negatively impact ecological health and the well-being of local communities over time.

Also, restoration efforts can be affected by drivers arising from a larger managerial and spatial scale that surpasses the local scale. For instance, community-based restoration activities can facilitate sustainable natural resource management on a local scale (e.g. those integrating rangeland restoration, biodiversity conservation, and livelihoods) but can be influenced by external drivers such as climate change, population growth, migration, fishing industries, and land subdivision authorised by government (e.g. Chap. 3). Even longstanding local restoration initiatives may face new threats, for instance in the sense that initiatives can be easily undone by government-funded large-scale development projects (e.g. Chaps. 10 and 13).

Furthermore, trade-offs between competing needs and interests extend beyond national boundaries. Certain selected endangered wildlife species (e.g. jaguars, sea turtles, and fish species) can be effectively conserved in one country through the enforcement of its national conservation policies, but if such species cross into another country that does not have an appropriate conservation policy, they may be subjected to exploitation like poaching. Thus, not only national but also regional and international responses or measures are important to support local restoration initiatives and intervene in external factors that have adverse effects on such local efforts (e.g. commercial fishing, plantation expansion, and perverse financial incentives). Here, different priorities across multiple sectors even at a certain governance level (i.e. lack of policy coherence) can be a barrier to a concerted restoration effort, confusing stakeholders with discrepant or contradictory policies and associated measures (e.g. food production policy vs. nature conservation policy) or sometimes having inconsistent financial support (e.g. budgeting conservation measures vs. incentivising farming practices that are not necessarily environmentally friendly).

Financial Sustainability and Adaptability

As mentioned above, restoration initiatives require new resources and capacities, which could be beyond what the local communities possess. Thus, external support (e.g. donor financing, resource mobilisation, and technology transfer) is often needed to initiate restoration activities, enhance local capacities, or help the government with reducing harmful incentives. However, external support could give rise to financial dependency or even trade-offs (e.g. ecotourism development leading to adverse environmental outcomes), whereby local communities may be confronted with new challenges derived from external factors (e.g. climate change and global trade). Therefore, financial sustainability is crucial to ensure good ideas and practices of natural resource use and management continue even without external funding.

Related to this, project ownership and the adaptive capacities of locals need to be fostered and continuously enhanced to address negative trade-offs and build resilience against changes or shocks. To do so, a post-project sustainability plan (e.g. strategic capacity development, empowerment, public-private partnership, networking, and mechanisms for incentives) should be developed and built into a restoration initiative. In addition, some incentive mechanisms (e.g. carbon credits, payments for ecosystem services (PES), awards to praise individuals or groups engaging in conservation, and farmers' rights to traditional crop varieties) help to ensure financial sustainability and facilitate long-term adaptive management, but may require the establishment of a new system (e.g. credit trading systems or markets, regulatory or administrative authorities). It is noteworthy that restoration activities conducted as part of regular production activities rather than as a formal project could also render valuable contributions to ecosystem restoration through the regenerative effects of SEPLS management (e.g. Chap. 7).

14.3.2 Way Forward: Opportunities for Synergistic Restoration

Landscape approaches offer an integrated scheme to collectively overcome the restoration challenges across different stakeholders, sectors, and levels, and to identify opportunities for synergies towards concerted efforts for restoration and sustainable development. Ways to exploit such opportunities lie in (1) multi-stakeholder participation and involvement, and (2) multi-lateral frameworks and coordination across different sectors and governance levels.

Multi-stakeholder Participation and Involvement
Bringing together multiple stakeholders on a common platform allows them to negotiate different needs and interests, share knowledge and learn from each other, be motivated to take action and mobilise resources, and finally collaborate on restoration. These stakeholders range from IPLCs, youth, women, and the private sector to government, scientists, and other experts. Each of them plays a key role in planning and implementing restoration efforts:

- *IPLCs*: Involvement of IPLCs in negotiation and decision-making for restoration is essential so that they can voice their needs and interests in regard to their rights and access to natural resource use as well as the environmental and socio-economic outcomes. This helps to ensure their livelihoods and well-being are secured and improved, while finding win-win solutions on the ground to attain multiple objectives (e.g. development of bio-cultural community protocols for livelihoods and conservation). It would also allow for the application of ILK to restoration and facilitate adaptive co-management based on lessons and experiences. Recording and documenting longstanding wise-use practices and ILK could be promoted and supported in consideration of the prehistoric role of indigenous peoples in managing SEPLS (e.g. ILK banks and global ILK support system). Yet, their intellectual property rights should be ensured, whereby cautious reflections are made for their participation in regard to time sensitivity (e.g. considering the time constraints of the locals for participation), FPIC, equity, equal opportunities, and gender equality (e.g. Chaps. 10 and 13).

- *Youth and women*: For sustainability of restoration, youth involvement is critical so that efforts continuously benefit future generations. Particularly in Africa where youth (aged 18–35 years) account for 60–70% of the population (AfDB et al. 2017), their engagement in restoration has a great potential for significance. Also, involvement of women and explicit recognition of their roles in SEPLS management promotes gender parity and extends the project scale to include diverse perspectives in the restoration efforts. For instance, the involvement of women in the development of restoration activities at the Mole Ecological Area, Ghana, resulted in women's effective engagement in managing a nursery that supplies seedlings of economically-important trees (e.g. *Parkia biglobosa* and *Vitellaria paradoxa*). These trees can be used for multiple purposes significant to the women, and thus were chosen as seedlings for restoring the degraded areas (Chap. 2). Participation by youth and women in restoration can be realised through unique and innovative approaches that can attract their attention, curiosity, and interest, while reflecting their specific needs. Such approaches can be applied to incentive creation (e.g. employment), communications (e.g. modern information technologies, social media, and traditional communication channels like community radio), and awareness raising and empowerment (e.g. dance, songs, street theatre, and poems) (see Chaps. 10, 12, and 13).
- *Private sector*: Involvement of the private sector (e.g. investors and private companies) in dialogues and decision-making can facilitate resource mobilisation and project implementation (e.g. Chap. 4), and enhance the financial sustainability of restoration, for instance, through public–private co-financing, payment for ecosystem services (PES) at the corporate level (linked to ecological footprint), and other forms of public–private partnerships. Impact investments may be linked to the products yielding from restored land and sea areas.
- *Government and administrative authorities*: Government engagement in restoration (particularly at the national level) allows for policymaking and implementation to address cross-boundary issues and deal with external forces (e.g. migration, climate change, and large-scale commercial fishing) (e.g. Chap. 3). At the same time, local authorities or traditional community-level institutions (e.g. district assemblies) can facilitate long-term bottom-up approaches to awareness raising, empowerment, resource mobilisation, and adaptive co-management for synergistic activities with keen attention to the well-being of local communities (e.g. taking advantage of interconnections between culture, ILK, ecotourism, climate action, agrobiodiversity, food security, and ecosystem restoration). For instance, Community Resource Management Committees (CRMCs) have served as a key local institution to facilitate mangrove restoration at two Ramsar sites in Ghana, which has contributed to both the conservation of aquatic biodiversity (e.g. birds and marine turtles) and the enhancement of community livelihoods through recruiting fisherfolks and promoting the use of fuel-efficient stoves by women for smoking fish (see Chaps. 9 and 13).

- *NGOs, scientists, and other experts*: Experts and professionals in NGOs, academia, and other organisations often serve as bridging stakeholders or facilitators to enable consultation and promote long-term cooperation across multiple stakeholders. Simultaneously, they may render their expertise in science, awareness raising, and community education, among others, to reinforce capacity development and resource mobilisation or develop new technological approaches to restoration interventions, monitoring, and evaluation. For instance, NGOs often play an indispensable role in bridging a gap between academia and local communities by strengthening mutual understanding and assisting in linking scientific knowledge with ILK (e.g. Chaps. 12 and 13).

Multi-lateral Frameworks and Coordination

To trigger and sustain a restoration initiative for sustainable development, the stakeholders need to feel motivated, agree on or if not negotiate their roles and responsibilities, collectively develop a plan, and collaborate for long-term adaptive co-management of natural resources. This process should be multi-lateral, iterative, and inclusive, and needs to be navigated by communicating and interacting with the stakeholders horizontally and vertically across different sectors and levels. For this to be achieved, the following three steps, which are not mutually exclusive, could be repeated and modified throughout the planning cycle:

- *Start from a landscape or seascape scale*:

  From the perspective of a landscape or seascape, policymakers and practitioners can find context-specific issues, learn about relevant stakeholders and how they connect to each other, and select appropriate methods and approaches that suit a certain place for restoration. Communication with local stakeholders allows for a better understanding of the locals' everyday practices and associated value perspectives as well as potential opportunities for restoration and threats to sustainable practices. This helps policymakers and practitioners determine what should not be left out as critical elements and what is feasible for restoration. At the same time, such communication helps the locals to recognise the problems or threats to biodiversity and ecosystems that consequently affect their livelihoods and well-being (e.g. climate change impacts), and thus the importance of and need for restoration. As such, they are more motivated to take action and engage in restoration and more sustainable natural resource management (e.g. Chap. 4). This land/seascape-scale consultation can facilitate capacity development and resilience building in SEPLS, which can further help to ensure financial sustainability and long-term adaptive co-management of natural resources.
- *Promote peer learning and knowledge sharing*:

  A platform or network for peer learning and knowledge sharing can serve to make a land/seascape-level restoration effort more effective and significant and help to upscale such an initiative for broader impacts. Such learning facilitates addressing multi-dimensional problems, and identifying and developing integrated solutions for restoration. This is because on the one hand, the factors associated with ecosystem degradation cannot be capsulated within the landscape or seascape scale but are interlinked to internal and external drivers possibly leading to extensive consequences beyond a certain spatial and/or temporal scale. On the other hand, good practices can be replicated or adapted even in different contexts to address common challenges, where lessons learnt from a land/seascape-scale initiative or local solutions can inform decision-making and actions by other stakeholders.

A common platform for learning and sharing enables the stakeholders to conduct comparative analyses (which help to better address challenges and explore opportunities by identifying both similar and distinctive drivers across different landscapes and seascapes), learn lessons from not only successes but also failures (which tend to be underreported), raise awareness and understanding among diverse stakeholders, and tap into different resources and capacities for restoration (Chaps. 11 and 13). Such a platform could be held either online or physically to share and learn lessons among diverse stakeholders (e.g. local, regional, national, and international) and make different kinds of knowledge accessible and available to other users if appropriate—though the intellectual property rights of knowledge holders need to be carefully attended.

• *Institutionalise local solutions into coherent policies and frameworks*:

To move towards more sustainable futures, local restoration efforts need to be institutionalised for systemic change in human–nature interactions. Restoration initiatives on the ground can be incorporated (i.e. linked, upscaled, or mainstreamed) into higher-level plans and strategies in a coherent manner (e.g. policy frameworks across local, regional, and national levels). In this regard, a customary or traditional local governance system (which anchors SEPLS management) plays an essential role in steering restoration initiatives, while being coordinated with government institutions to allow for synergistic policymaking and implementation. Here, upstream and downstream connections should also be recognised along with clear, specific roles and responsibilities among stakeholders.

To develop such a coherent and comprehensive governance structure, global policy frameworks [e.g. UN Decade on Ecosystem Restoration, UN Decade on Ocean Science, UN Decade on Family Farming, other effective area-based conservation measures (OECMs)] can be drawn on to make local contributions more visible and recognisable at multiple levels towards achieving global goals for sustainability. Particularly for production activities (e.g. food production), supply chain frameworks can also be applied to identify interventions across different stages of value chains (e.g. production, distribution, retailing, and consumption) and integrate local restoration actions, from SEPLS management perspectives, into broader economic institutions where financial flows could be calibrated.

Furthermore, building on certain integrative concepts for sustainability (e.g. SEPLS, nature-based solutions, and ecosystem approach), multi-stakeholder networks, particularly with involvement of policymakers (e.g. IPSI), can be created and fostered at the international level (wherein national and regional coordination can help strengthen the network as appropriate). This would serve to heighten the pride and motivation of local stakeholders, solicit wider interest and attention to local actions, and advocate for broader support. It would simultaneously help to raise local capacities for long-term commitment to meeting global goals for restoration and sustainable development.

14.4   Conclusion

Ecosystem restoration entails rehabilitating and rejuvenating ecosystems to ensure that their functional integrity is restored and sustained. Of special interest to this volume are also the implications this would have on biodiversity and the influence of (and on) cultural practices of people in SEPLS on achieving this goal. While engaging in activities that prevent, halt, and reverse degradation, it is important to address various drivers (whether natural or anthropogenic including economic, political, or social). Addressing these drivers involves attaining the cooperation of different actors and stakeholders with multiple priorities, responsibilities, and decision-making capacities. At the SEPLS level, this would typically involve engaging with local communities (and indigenous peoples), government bodies, resource management authorities, industrial bodies, and researchers.

The experiences from across different case studies illustrate that successful restoration activities are dependent on identifying potential trade-offs arising from various activities in a landscape or seascape. These trade-offs could occur within or between different scales of implementation, often privileging the interests of some dominant actors. Lessons from the case studies show that flexible, adaptive management strategies, which are developed by inclusive, participatory methods with a clear purpose building on existing assets (such as natural resources, knowledge and sustainable resource use practices, and human resources), are most likely to lead to successful restoration outcomes. Innovative use of digital technology to share information, raise awareness, and access knowledge and external networks complements this approach along with a variety of incentives from carbon credits, and awards for good practice, among many others.

Nevertheless, operationalising restoration activities are affected by various challenges that range from political will and capacity asymmetries relating to information and resources, to power asymmetries between sectors and actors, among others. Addressing these requires synergising across various sectoral initiatives, enabling deliberative processes, developing capacities related to understanding social-ecological underpinnings of restoration activities, and further integrating land/sea-scape-level solutions to policy interventions. Leveraging existing sectoral policy initiatives with potential for collaboration with other sectors allows for more effective use of various resources (financial, human, natural), enhances better understanding of the potential outcomes of different interventions, and fosters better outcomes for people and nature.

The UN Decade on Ecosystem Restoration and similar efforts provide a much-needed political impetus to advance such approaches. Hopefully, these initiatives will translate into focused investments (financial, technical, and human resources) in capacity development and research, and training and peer learning efforts that are mindful of advancing social-ecological resilience and enable restoration goals to be achieved.

References

AfDB, OECD, UNDP (2017) African economic outlook 2017: entrepreneurship and industrialisation. OECD Publishing, Paris. https://doi.org/10.1787/aeo-2017-en

Alejos SF, Pajaro M, Raquino M, Stuart A, Watts P (2021) Agricultural biodiversity and coastal food systems: a socio-ecological and trans-ecosystem case study in Aurora Province, Philippines. Asian J Agric Dev 18(1362-2022-017):74–84

Dublin DR, Natori Y (2020) Community-based project assessment using the indicators of resilience in SEPLS: lessons from the GEF-Satoyama Project. Curr Res Environ Sustain 2:100016. https://doi.org/10.1016/j.crsust.2020.100016

Dunbar W, Subramanian SM, Matsumoto I, Natori Y, Dublin D, Bergamini N, Mijatovic D, Álvarez AG, Yiu E, Ichikawa K, Morimoto Y, Halewood M, Maundu P, Salvemini D, Tschenscher T, Mock G (2020) Lessons learned from application of the "Indicators of resilience in socio-ecological production landscapes and seascapes (SEPLS)" under the Satoyama initiative. In: Managing socio-ecological production landscapes and seascapes for sustainable communities in Asia: mapping and navigating stakeholders, policy and action, pp 93–116

Edake S, Sethi P, Lele Y (2019) Mainstreaming Community-Conserved Areas (CCAs) for biodiversity conservation in SEPLS-A case study from Nagaland, India. In: UNU-IAS, IGES (ed) Understanding the multiple values associated with sustainable use in socio-ecological production landscapes and seascapes (Satoyama initiative thematic review), vol 5. United Nations University Institute for the Advanced Study of Sustainability, Tokyo, pp 169–179

FAO, IUCN, CEM, SER (2021) Principles for ecosystem restoration to guide the United Nations decade 2021–2030

FAO, UNEP (2022) Global indicators for monitoring: a contribution to the UN decade on ecosystem restoration. https://doi.org/10.4060/cb9982en

Filbee-Dexter K, Smajdor A (2019) Ethics of assisted evolution in marine conservation. Front Mar Sci 6. https://doi.org/10.3389/fmars.2019.00020

Florentina-Cristina M, Sirodoev I, George M, Zamfir D, Schvab A, Stoica I, Paraschiv M, Irina S, Cercleux A-L, Vaidianu N, Ianos I (2017) The "Văcărești Lake" protected area, a neverending debatable issue? Carpathian J Earth Environ Sci 12:463–472

Hopfensperger KN, Engelhardt KAM, Seagle SW (2007) Ecological feasibility studies in restoration decision making. Environ Manag 39(6):843–852. https://doi.org/10.1007/s00267-005-0388-7

Iftekhar MS, Polyakov M (2021) Economics of ecological restoration. Environ Sci. https://doi.org/10.1093/acrefore/9780199389414.013.751

Iftekhar MS, Polyakov M, Ansell D, Gibson F, Kay GM (2017) How economics can further the success of ecological restoration. Conserv Biol 31(2):261–268

Israel BA, Schulz AJ, Parker EA, Becker AB (1998) Review of community-based research: assessing partnership approaches to improve public health. Annu Rev Public Health 19:173–202. https://doi.org/10.1146/annurev.publhealth.19.1.173

Karimova PG, Yan SY, Lee KC (2022) SEPLS well-being as a vision: co-managing for diversity, connectivity, and adaptive capacity in Xinshe Village, Hualien County, Chinese Taipei. In:

Nishi M, Subramanian SM, Gupta H (eds) Biodiversity-health-sustainability nexus in socio-ecological production landscapes and seascapes (SEPLS) (Satoyama initiative thematic review), vol 7. Springer, Cham, pp 61–88

Keahey J (2021) Sustainable development and participatory action research: a systematic review. Syst Pract Action Res 34(3):291–306. https://doi.org/10.1007/s11213-020-09535-8

Ken S, Entani T, Tsusaka TW, Sasaki N (2020) Effect of REDD+ projects on local livelihood assets in Keo Seima and Oddar Meanchey, Cambodia. Heliyon 6. https://doi.org/10.1016/j.heliyon.2020.e03802

Lax J, Krug J (2013) Livelihood assessment: a participatory tool for natural resource dependent communities (Issue 7). https://nbn-resolving.de/urn:nbn:de:gbv:253-201308-dn052272-8

Maxwell SM, Hazen EL, Lewison RL, Dunn DC, Bailey H, Bograd SJ, Briscoe DK, Fossette S, Hobday AJ, Bennett M, Benson S, Caldwell MR, Costa DP, Dewar H, Eguchi T, Hazen L, Kohin S, Sippel T, Crowder LB (2015) Dynamic ocean management: Defining and conceptualising real-time management of the ocean. Mar Policy 58:42–50. https://doi.org/10.1016/j.marpol.2015.03.014

McDonald T, Gann GD, Jonson J, Dixon KW (2016) International standards for the practice of ecological restoration—including principles and key concepts. Society for Ecological Restoration, Washington, DC

Menhinick EF (1964) A comparison of some species-individuals diversity indices applied to samples of field insects. Ecology 45(4):859–861. https://doi.org/10.1038/163688a0

NCC (2013) The state of natural capital: towards a framework for measurement and valuation (Issue March). http://www.defra.gov.uk/naturalcapitalcommittee/

Orlando LF, DePinto AJ, Wallace KJ (2022) Ecohealth villages: a framework for an ecosystem approach to health in human settlements. Sustainability 14(12):7053. https://doi.org/10.3390/su14127053

Peck S (1998) Planning for biodiversity: issues and examples. Island Press, Washington, DC

Pielou E (1969) An introduction to mathematical ecology. Science 169(3940):43–44. https://doi.org/10.1126/science.169.3940.43-a

Schultz ET, Johnston RJ, Segerson K, Besedin EY (2012) Integrating ecology and economics for restoration: using ecological indicators in valuation of ecosystem services. Restor Ecol 20(3):304–310. https://doi.org/10.1111/j.1526-100X.2011.00854.x

Simpson E (1949) Measurement of diversity. Nature 163(1943):688. https://doi.org/10.1038/163688a0

Supriatna J (2018) Biodiversity indexes: value and evaluation purposes. E3S Web Conf 48:01001. https://doi.org/10.1051/e3sconf/20184801001

Tengö M, Brondizio ES, Elmqvist T (2014) Connecting diverse knowledge systems for enhanced ecosystem governance: the multiple evidence base approach. Ambio 43:579–591. https://doi.org/10.1007/s13280-014-0501-3

Turner MG, Gardner RH, O'Neill RV (2001) Landscape ecology in theory and practice: pattern and process. Springer, Cham

Villarreal-Rosas J, Vogl AL, Sonter LJ, Possingham HP, Rhodes JR (2021) Trade-offs between efficiency, equality and equity in restoration for flood protection. Environ Res Lett 17(1):014001. https://doi.org/10.1088/1748-9326/ac3797

Watts PD, Pajaro MG (2014) Collaborative Philippine-Canadian action cycles for strategic international coastal ecohealth. Can J Action Res 15(1):3–21

Watts P, Custer B, Yi Z, Ontiri E, Pajaro MA (2015) Yin-Yang approach to education policy regarding health and the environment: early-careerists' image of the future and priority programmes. (United Nations). Nat Res Forum 39:202–213. https://doi.org/10.1111/1477-8947.12083

Watts PD, Pajaro MG, Raquino MR, Añabieza JM (2021) Philippine fisherfolk: sustainable community development action research and reflexive education. Local Dev Soc 3:1–19. https://doi.org/10.1080/26883597.2021.1952847

Wortley L, Hero JM, Howes M (2013) Evaluating ecological restoration success: a review of the literature. Restor Ecol 21(5):537–543. https://doi.org/10.1111/rec.12028

Zhai L, Cheng S, Sang H, Xie W, Gan L (2022) Remote sensing evaluation of ecological restoration engineering effect: a case study of the Yongding River Watershed. Chin Ecol Eng 182 (June):106724. https://doi.org/10.1016/j.ecoleng.2022.106724

Zhang H, Zhao Y, Pedersen J (2020) Capital assets framework for analysing household vulnerability during disaster. Disasters 44(4):687–707. https://doi.org/10.1111/disa.12393

The opinions expressed in this chapter are those of the author(s) and do not necessarily reflect the views of UNU-IAS, its Board of Directors, or the countries they represent.

Open Access This chapter is licenced under the terms of the Creative Commons Attribution-NonCommercial-ShareAlike 3.0 IGO licence (http://creativecommons.org/licenses/by-nc-sa/3.0/igo/), which permits any noncommercial use, sharing, adaptation, distribution and reproduction in any medium or format, as long as you give appropriate credit to UNU-IAS, provide a link to the Creative Commons licence and indicate if changes were made. If you remix, transform, or build upon this book or a part thereof, you must distribute your contributions under the same licence as the original. The use of the UNU-IAS name and logo, shall be subject to a separate written licence agreement between UNU-IAS and the user and is not authorised as part of this CC BY-NC-SA 3.0 IGO licence. Note that the link provided above includes additional terms and conditions of the licence.

The images or other third party material in this chapter are included in the chapter's Creative Commons licence, unless indicated otherwise in a credit line to the material. If material is not included in the chapter's Creative Commons licence and your intended use is not permitted by statutory regulation or exceeds the permitted use, you will need to obtain permission directly from the copyright holder.

Printed in the United States
by Baker & Taylor Publisher Services